# Microbiology for the Analytical Chemist

# Microbiology for the Analytical Chemist

**R.K. Dart**
*Loughborough University*

THE ROYAL SOCIETY OF CHEMISTRY
Information Services

ISBN 0-85404-524-4

A catalogue record of this book is available from the British Library



Published by The Royal Society of Chemistry,
Thomas Graham House, Science Park, Milton Road, Cambridge CB4 4WF, UK

Typeset by Computape (Pickering) Ltd, Pickering, North Yorkshire, UK
Printed by Redwood Books Ltd., Trowbridge, Wiltshire.

# Preface

The microbiological problem facing analytical chemists is that when a sample is compromised microbiologically, it is only a matter of time before it is compromised chemically/biochemically. In the case of heavy microbial contamination, the period of time before a sample becomes compromised chemically, may only be a few hours.

The majority of problems in analytical chemistry caused by micro-organisms take the form of four questions. These are:

(a) Is the sample sterile? Technically, sterility is an absolute value, and there are no degrees of sterility, although as will be seen, this question is not as straightforward as it seems. The answer to the question should therefore be either 'Yes' or 'No'. If the answer is 'Yes', then there is generally no further microbiological problem for the analytical chemist. If the answer is 'No', then a number of further questions arise.
(b) How many micro-organisms are present?
(c) What sort are they?
(d) What effect do they have on the product/sample?

It is the answers to these four questions which this short book addresses.

It should be appreciated that this is not a comprehensive microbiology text, but is confined to those micro-organisms of medicinal and industrial importance, which experience suggests will be found in situations encountered by analytical chemists. This means that organisms labelled medicinal in the previous sentence, generally will be those causing food poisoning, whilst those labelled industrial will, broadly speaking, be those causing spoilage.

# Contents

CHAPTER 1

# Introduction

## 1 THE MICROBIAL CELL

Microbiology can be defined as the study of organisms that are too small to be clearly visible to the naked eye. This will include all organisms with a diameter of less than approximately 1 mm.

Micro-organisms are widely distributed between different taxonomic groups and include bacteria, protozoa, fungi, and algae. Viruses, although frequently considered in microbiology text books, are not cells.

The discovery of micro-organisms caused numerous problems relating to the placing of these species into the traditional plant and animal kingdoms, and in the mid-19th century a third kingdom, the *Protista* was suggested to include the protozoans, fungi, algae, and bacteria.

The development of electron microscopy showed that there was a major dichotomy between the various groups relating to the internal structure and organisation of cells. Two very different types of cells were discovered, the small, relatively simple procaryotic cell, and the more complex eucaryotic cell which is usually considerably larger.

### 1.1 Procaryotic Cells

The bacteria are procaryotic cells. This group also includes the organisms which used to be known as the 'Blue–green Algae' or Cyanophyceae, and are now known as the Cyanobacteria.

The vast majority of procaryotic organisms are unicellular, although multicellular bacteria forming filaments are also found. A few types are more complex structurally and may produce fruiting bodies, appendages, or stalks, although these would not generally be found in a typical analytical laboratory. Bacteria forming filaments frequently show a marked tendency to break up into individual cells. Figure 1.1 shows a typical procaryotic cell.

Procaryotic cells are surrounded by a cell membrane, a thin flexible sheet composed of protein and lipid. This has an important function in regulating the molecules which can pass into and out of the cell.

In the great majority of cases, there is a rigid cell wall outside the membrane.

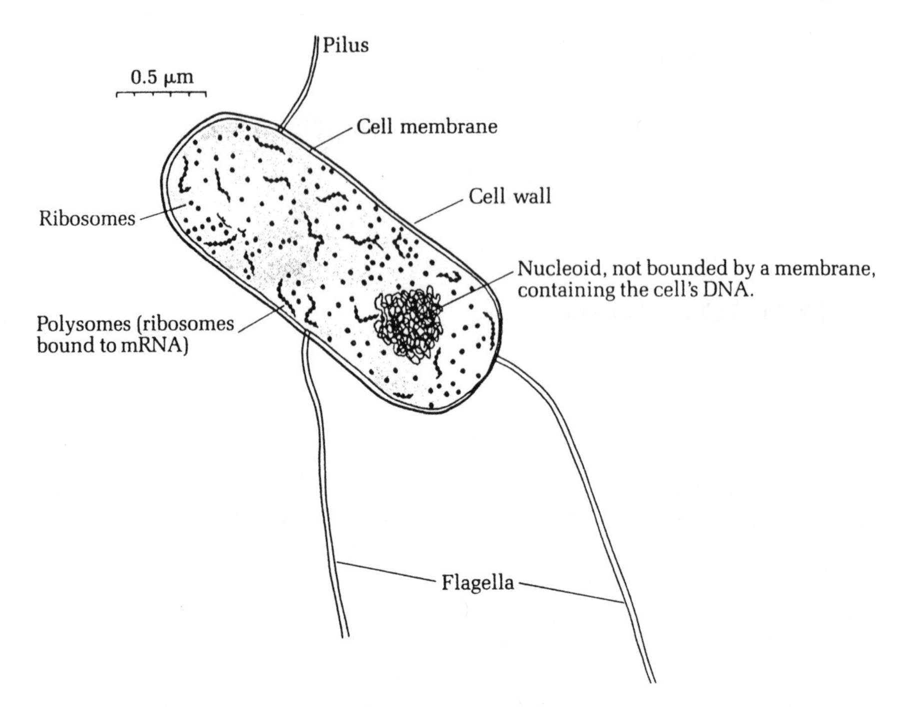


**Figure 1.1** *Diagram showing a typical procaryotic cell*

This cell wall has a chemistry which is unique to the procaryotic cell, and contains a number of compounds which are not found in the eucaryotes. These include peptidoglycan (murein), which is a repeating disaccharide containing the amino sugar, *N*-acetylmuramic acid, linked to a peptide containing D-amino acids, and diamino acids such as diaminopimelic acid. This cell wall chemistry is an important linking factor between the true bacteria and the blue–green algae. A few groups such as Mycoplasma do not possess a cell wall.

In the procaryotes, there is no discrete nucleus surrounded by a nuclear membrane, and the deoxyribonucleic acid (DNA) is not associated with the basic proteins known as histones. The DNA consists of covalently closed circles.

There are no organelles such as mitochondria present within the cell. Flagella are present in the case of motile cells; the position of these varies from species to species and may be used diagnostically. Certain species may contain spores which are highly resistant to heat and a variety of chemicals. A number of species are capable of photosynthesis, that is they can convert solar energy into chemical energy. Those that do carry out photosynthesis have chromatophores which are extensions of the cell membrane, and which are morphologically distinct from the chloroplast of the eucaryotic higher plant (see later).

There is also a considerable difference in the ribosome structure of procaryotic and eucaryotic cells. Ribosomes are the structures upon which proteins are synthesised. Procaryotic cells possess ribosomes which are 70 S in size, whereas eucaryotes have 80 S ribosomes.

## 1.2 Eucaryotic Cells

Organisms containing eucaryotic cells include animals, plants, fungi, algae, and protozoa. Figure 1.2 shows typical animal and plant eucaryotic cells.

A cell membrane consisting of a protein and lipid mixture is present, but the eucaryotic cell membrane contains significant quantities of steroids, which are not found in most procaryotic cells. There may or may not be a cell wall outside the membrane depending on the type of cell.

Eucaryotic cells possess a nucleus, surrounded by a nuclear membrane which delineates the nucleus from the rest of the cell. This nucleus contains the majority of the cell DNA and is associated with histones. The nuclear DNA is linear. The nucleus is also the centre for the synthesis of ribonucleic acid (RNA).

All material inside the cell membrane, but outside the nuclear membrane, is known as the cytoplasm. Other organelles are found in the cytoplasm of the eucaryotic cell. These include the mitochondria, which are the powerhouse of the eucaryotic cell producing much of the energy in the form of adenosine triphosphate (ATP). Chloroplasts are found in photosynthetic species such as plants and algae. Neither of these structures is found in procaryotes. Various other organelles such as a nucleolus, centrioles, and a Golgi apparatus are also found. There may be one or more cilia or flagella in the case of motile cells, but these differ considerably from the flagella of bacterial cells.

Eucaryotic micro-organisms of importance to the analytical chemist include the fungi, protozoa and algae.

## 1.3 Nomenclature and Taxonomy

Micro-organisms are named using a binomial system. Every species has two latinised names, the first being the generic name and the second the specific name which identifies the organism within the genus. The first letter of the generic name is in capitals and the whole name is either italicised or underlined. In situations where no ambiguity can arise only the first letter or a shortened version of the generic name may be used, although the full name should always be used the first time it arises.

Nomenclature may cause problems. Occasionally, newly isolated organisms may not be recognised for what they are, and are given a new name. This can cause considerable confusion when the organism is finally correctly identified.

Taxonomy is the science of classification and has several functions. The first is to describe the species which is the basic taxonomic unit. The second is to catalogue these species into some arrangement enabling the relationships between species to be recognised. A third practical aspect is identification, that is, the matching of an unknown organism with a known species.

The division between species at the plant and animal levels is generally relatively easy on the basis of their morphology (appearance), although problems can arise in distinguishing between the simpler algae and protozoa. This however is not true with bacteria as the range of distinguishable shapes and sizes is too small.

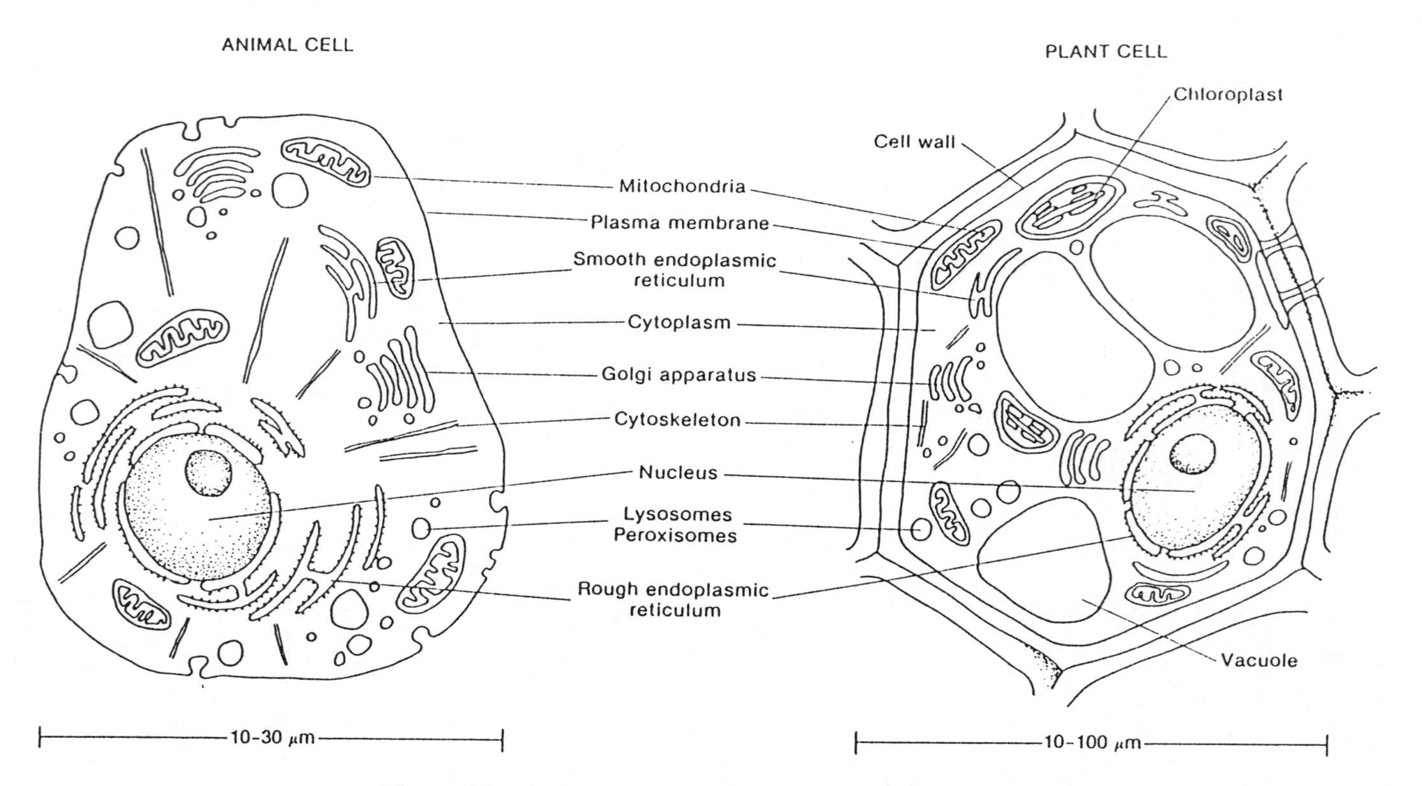


**Figure 1.2** *Diagram showing typical animal and plant eucaryotic cells*

Bacteria are therefore classified and identified into strains on the basis of a number of tests (*e.g.* morphological, staining, biochemical, serological). A species may then be defined as a cluster of strains showing a high degree of similarity, but differing from other clusters of strains (species) in a number of characteristics.

In taxonomy, the strains are grouped into individual species which are then grouped into progressively higher series of categories. In order, these are the genus, which is a collection of species, the family, the order, the class and the phylum. This is known as a hierarchical classification, as each ascending category unites a larger number of taxonomic groups on the basis of a smaller number of common properties. This hierarchical type of approach reflects to a considerable extent the evolutionary relationships of the organisms involved. In bacteriology, the two important groupings are the species and the genus. The various other hierarchical groups are not of great practical significance.

The use of various characteristics carries no guarantee that the tests being used are significant. The classification of bacteria therefore tends to be heavily biased towards the practical aspects of microbiology, such as the identification of medically and industrially important species.

The hierarchical type of approach to practical bacterial identification makes use of a number of characteristics which are given differential or weighted values. Tests are carried out in a pre-determined order designed to confirm or eliminate various possibilities. By following this pre-determined order, the experienced microbiologist is frequently able to reduce the possibilities to a small number of groups relatively quickly. In the majority of cases, characteristics would be examined in the following approximate order: source of the organism, colonial appearance on the growth medium, staining followed by microscopic examination, a variety of biochemical tests, and finally serological tests.

Other types of taxonomy are also found. Numerical classification or Adansonian classification is based on the quantification of similarities and differences, with no weighting being given to any one character. It should therefore be possible to quantify the relationships between cells on the basis of the number of similarities they share, relative to the number examined. The method is basically a cluster analysis in $n$-dimensional space, where $n$ is the number of variables examined. The clusters can then be related to each other by means of a dendrogram. In this type of classification, individual characters gain an importance not given to them in hierarchical taxonomy. The great advantage of this method is that it requires the determination of a large number of characteristics. The method also requires the use of a computerised data base and is now used as the basis of a number of rapid identification systems, *e.g.* the API system.

Irrespective of the type of classification system used, practical classification and identification on a day to day basis must, of necessity, be based on characteristics which are easy and simple to observe and measure. One obvious example is motility, which depends on flagella. Motility/non-motility is a useful differential character, which is easy to observe using simple techniques and is therefore widely used in classification. However, the number and position of the flagella, which may be polar or peritrichous (all round the cell), is also a differential character. To observe these reliably requires an electron microscope, which is not a generally available laboratory tool, and therefore this character is not used in routine identification.

Other classification methods, such as the analysis of guanine and cytosine (GC) ratios in DNA, and DNA homology have no place as yet in the routine testing of the analytical laboratory.

From the practical point, the most important classification relates to safety, and several categories of micro-organisms can be defined depending on how dangerous they are and the level of containment required.

Group 1: Organisms unlikely to cause disease in humans.

Group 2: Organisms that may cause human disease and may be a hazard to laboratory workers, but are unlikely to spread in the community. Laboratory exposure rarely causes infection and effective prophylaxis and/or treatment is usually available.

Group 3: Organisms that may cause severe human disease and which present a serious hazard to laboratory workers. There may be a risk of spread in the community but there is usually effective prophylaxis and/or treatment available.

Group 4: Organisms that cause severe human diseases and are a severe hazard to laboratory workers. There is a high risk of spread in the community, and there is usually no effective prophylaxis and/or treatment.

For most practical purposes, the organisms found in an analytical laboratory will usually fall into Groups 2 or 3, and therefore laboratories should be of a sufficiently high standard to handle Group 3 type organisms.

## 2 EUCARYOTES

The eucaryotic micro-organisms include the protozoa, fungi and algae. The inclusion of some of the more highly specialised members into these three groups can be easily recognised. However, for some of the simpler ones such divisions may be difficult to recognise, and transition between algae and protozoa is common.

### 2.1 Algae

The algae are organisms possessing chloroplasts, carrying out photosynthesis and producing oxygen. They are a highly diverse group which can be divided into a number of classes. Many algae are unicellular micro-organisms, but they also contain giant multicellular species such as the kelps which may grow to a length of 50 m. Many of the unicellular species are motile by means of one or more flagella. Their main interest to the analytical chemist is the ability of some groups of unicellular planktonic species to cause the so-called 'red tides', and produce potent phytotoxins causing poisoning of shellfish and other species.

## 2.2 Protozoa

The protozoa are a very diverse group of unicellular organisms lacking in photosynthetic abilities. Most of them show no resemblance to algae, but some of the simpler species are obviously related to algae which have lost their photosynthetic apparatus. This loss would restrict the range of environments available to the species, and cause a series of evolutionary changes leading eventually to protozoans whose algal origins have become unrecognisable.

A number of classes of protozoans can be distinguished, and there are a number of species which parasitise man. Many of these are carried by insect vectors, causing diseases such as malaria and sleeping sickness (trypanosomiasis), and are beyond the scope of this book. Several species are water-borne or may be found on uncooked food contaminated by sewage, and when ingested can cause diseases such as amoebic dysentery. Many of these are tropical or sub-tropical, but one water-borne organism causing problems in the UK is a member of the genus Giardia.

## 2.3 Fungi

The fungi are also a non-photosynthetic group, the majority being highly specialised and well adapted to soil which is their main habitat. A few very primitive aquatic species, which are motile, show some resemblance to flagellated protozoans. Figure 1.3 shows a diagrammatic representation of a typical fungus.

Most fungi consist of a mycelium which is a highly branched series of tubes containing multiple nuclei. Individual branches are referred to as hyphae. Asexual reproduction usually occurs by means of spores pinched off at the tip of the hypha.

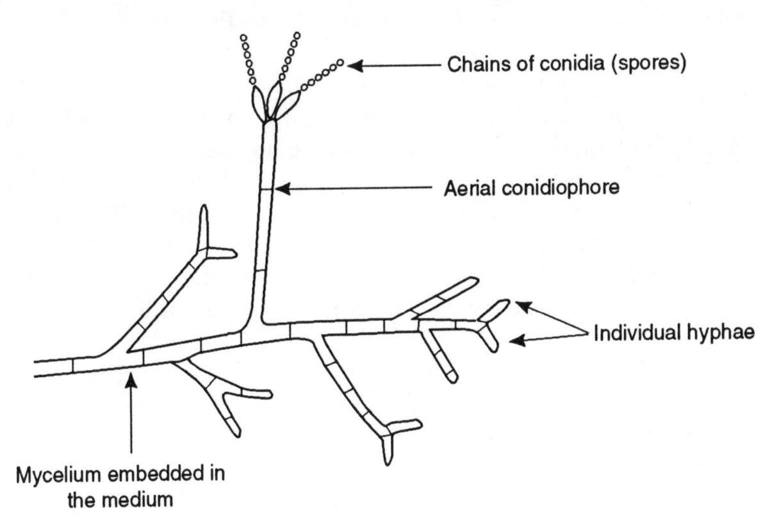


**Figure 1.3** *Diagrammatic representation of a typical fungus (*Penicillium, *a member of the Ascomycetes)*

The cell wall is composed mainly of a polymer of *N*-acetylglucosamine known as chitin, although there may also be limited amounts of cellulose present.

The fungi are generally sub-divided into four groups based on the structure of their sexual stages, although there is some controversy over the number. These are the Phycomycetes, Ascomycetes, Basidiomycetes, and Deuteromycetes (Fungi Imperfecti).

The Phycomycetes are the simplest group, although considerable differences can be found within them depending on whether the species being studied is aquatic or terrestrial. They have two common features which distinguish them from other fungi. The first is the asexual spores which are always endogenous, *i.e.* they are formed inside a saclike structure, whereas in other groups the asexual spores are exogenous or formed externally on the tips of the hyphae. Secondly the mycelium in the Phycomycetes has no cross walls (septa), whereas all other groups have cross walls. Phycomycetes commonly found in spoilage situations include species from the genera *Mucor* and *Rhizopus*.

Fungi with septa and external (exogenous) asexual spores are divided into Ascomycetes and Basidiomycetes on the basis of the development of their sexual spores. The structures involved, which are morphologically very different, are known as the asci (singular ascus) or basidia (basidium). Numerous fungi from the Ascomycetes are found in spoilage situations, where they may act either by toxin production or cause chemical damage leading to physical deterioration. Species from the genera *Penicillium*, *Aspergillus*, and *Fusarium* are very common in these situations.

It is obviously impossible to classify any fungus whose sexual stage is unknown, and these species are placed in the Fungi Imperfecti. This is essentially a holding group, and as the sexual stage of an organism is discovered it is transferred to one of the other groups, usually the Ascomycetes or Basidiomycetes. This can cause problems with organisms being found under two generic names, one relating to the perfect or sexual stage, and the other relating to the imperfect or asexual stage.

The basic structure amongst the Ascomycetes, Basidiomycetes and Fungi Imperfecti is the mycelium. There are, however, a number of fungi which have lost the mycelial growth characteristic and become unicellular. These unicellular fungi are known as yeasts. Typically, yeasts are oval cells which multiply by budding or fission. In yeasts that divide by budding, the parent cell forms a bud which enlarges until it is almost as big as the parent, nuclear division takes place and a cross wall is formed producing two cells (see Figure 1.4). Division by fission occurs when a cell grows, then lays down a cross wall to produce two equal sized cells (see Figure 1.5).

Although the yeasts are a relatively minor group numerically, they are important both industrially and medically. The majority do not live in the soil but have become adapted to environments with a high sugar content, such as fruit and honey, and although responsible for the production of alcoholic drinks such as wine, may also cause serious spoilage problems in a variety of fruit and sugar based products.

Some fungi are dimorphic and can exist in either the yeast or mycelial states,

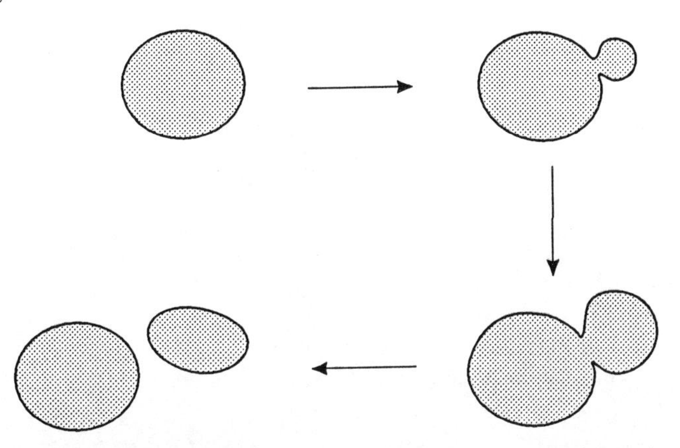

**Figure 1.4** *Budding of vegetative cells in yeast, e.g.* Saccharomyces, *in which the bud grows, eventually developing into a full sized cell*

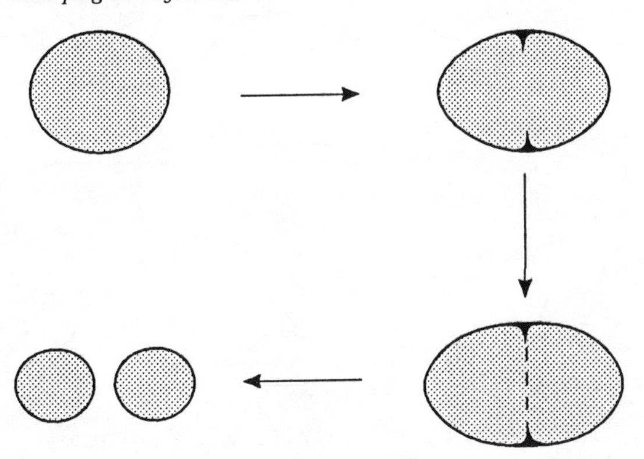

**Figure 1.5** *Fission in yeasts in which a parent cell divides giving two daughter cells of equal size,* e.g. Schizosaccharomyces

depending on the environmental conditions. Organisms capable of existing in dimorphic states are frequently pathogenic.

## 3 PROCARYOTES

The procaryotic group includes the Eubacteria and the Archaebacteria which differ markedly from the Eubacteria in their cell chemistry. Also included are the Cyanobacteria or Blue–green bacteria, which were formerly known as the Cyanophyceae or 'Blue–green algae'.

### 3.1 Eubacteria

The Eubacteria are a vast heterogeneous group of procaryotic cells which can be classified in numerous ways. These include their morphology and staining, their nutritional, oxygen, and temperature requirements, and a variety of other biochemical factors.

Cocci

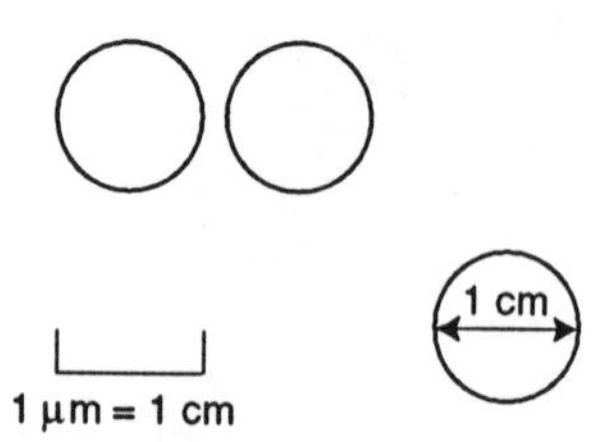


Cocci may be present as single cells, clumps, short or long chains

Rods

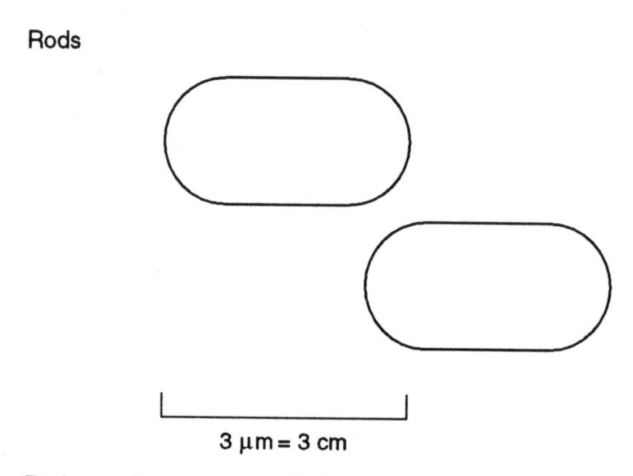


Rods may be present as single cells or short chains

**Figure 1.6** *Diagramatic representation showing cocci and rods*

Morphologically, several groups can be differentiated, although most of those of analytical importance are basically unicellular.

The cocci are spherical cells with a diameter of 1–2 µm. After division the cells may remain joined together to form pairs, chains, or clusters (see Figure 1.6). They include a number of spoilage and food poisoning organisms.

The rods are cylindrical bacteria up to 10 µm long and 1 µm in diameter. They may grow individually or in chains and may or may not be motile. If they are motile (see Figure 1.6), the number and position of flagella are variable, and in some cases may be used for diagnostic purposes. They include a large number of food poisoning and spoilage organisms.

The vibrios are short rod shaped organisms (cocco-bacilli) which are motile. They only consist of one group with a large polar flagellum and are commonly found in polluted water, especially in the tropics where they can cause cholera.

Some bacteria have a spiral cell structure. A number of these are pathogens found in water, and they are generally a difficult group to work with and isolate. These may show motility, but this is due to flexing of the cells and not the presence of flagella.

Bacteria forming a mycelium are also found. This group is known as the Actinomycetales and shows a wide divergence of shapes. The mycelium is much thinner than the fungal mycelium, and displays a marked tendency in some members of the group to break up into unicellular forms, which may be of unusual shape, sometimes referred to as Chinese lettering.

The Cyanobacteria are a group of bacteria which are structurally very diverse, ranging from unicellular species to filamentous ones which may be covered by a sheath. They all carry out photosynthesis and are differentiated from other photosynthetic bacteria by their range of pigments and their ability to form oxygen during photosynthesis. Many are capable of nitrogen fixation (*i.e.* conversion of gaseous nitrogen to organic nitrogen). Their nitrogen fixing ability has led to their use as a natural fertiliser in South East Asia. The main problem caused by this group is their ability to cause 'blooms' in water supplies at certain times of the year, leading to the total removal of oxygen from the water, death of other species and development of toxins in the water.

The eubacteria may be subdivided on the basis of a variety of reactions. Some of the more important ones include different staining reactions. The most important of these is the Gram stain which differentiates the Eubacteria into Gram-positive and -negative on the basis of their cell wall chemistry. Gram-positive bacteria have a thick cell wall of peptidoglycan, whereas Gram-negative bacteria have a much thinner layer. Other important stains used include the spore stain and acid-fast stain, both of which are used for the differentiation of some Gram-positive rods. Some books recommend the use of flagellar stains, but these require a high degree of technical competence to produce reliable results. Further sub-division of the bacteria makes wide use of growth on various selective and differential media, and also a number of biochemical tests. These will be discussed in detail later. These tests give rise to a considerable nomenclature based on a variety of properties. The various terms will be defined as they arise.

Several bacterial classification systems are known, but the one used throughout this book is that found in Bergey.[1,2] This reference exists as a four volume series,[1] and also a single volume version[2] which is adequate for most laboratory purposes.

### 3.2 Archaebacteria

Relatively little is known about this group of procaryotes, and from an industrial viewpoint the main interest in them is that they contain a group, known as the methanogens, which is able to form methane from carbon dioxide and acetic acid.

They can be divided into several groups, most of which are very difficult to grow and are generally considered to be so different to the Eubacteria that most authors place them in a separate kingdom.

### 3.3 Culture and Growth

Bacteria have a number of requirements for growth. These include sources of energy, carbon, nitrogen, phosphate, sulfate, and various trace elements. They may include vitamins in certain cases. They also require to be grown at the correct temperature, pH, osmotic pressure, and oxygen levels. There may be some overlap between carbon, nitrogen, and energy requirements. Thus, an amino acid may be an energy source as well as being a carbon and nitrogen source.

Bacteria may be divided into a number of groups depending on the types of energy and carbon sources they use. The basic divisions are into phototrophs using light as their energy source, and chemotrophs, which use organic chemicals as their energy source. Heterotrophism is the use of an organic source of carbon, and autotrophism is the use of an inorganic source ($CO_2$).

This gives four major groups:

(a) *Photoautotrophs* which use light as energy source and $CO_2$ as carbon source;
(b) *Chemoautotrophs* which oxidise reduced inorganic compounds such as $NH_4^+$, $NO_2$, $H_2$, $H_2S$, CO, *etc.* to obtain energy, and use $CO_2$ as carbon source;
(c) *Photoheterotrophs* which use light as energy source and some organic compound as carbon source;
(d) *Chemoheterotrophs* which use an organic compound as energy source and an organic compound as carbon source. In this case, the clear cut distinction between carbon and energy source is lost, as many chemoheterotrophs use the same compound for both purposes.

The vast majority of bacteria of industrial and medical importance are heterotrophic (chemoheterotrophs).

The terms obligate and facultative are also used; obligate organisms can only use that form of nutrition, whilst facultative can switch from one form to another. The most common example of facultative organisms is the algae, many of which are photoautotrophs in the light, but may be chemoheterotrophs in the dark.

Many bacteria have an aerobic form of metabolism utilising oxygen as the final electron acceptor and are referred to as aerobic. Some bacteria are able to carry out metabolism anaerobically (*i.e.* in the absence of oxygen), using some organic compound as final electron acceptor. These are known as anaerobic. Many bacteria are able to grow in either the presence or the absence of oxygen and are known as facultative, whilst a few which grow best in the presence of reduced levels of oxygen are referred to as microaerophilic.

Micro-organisms also require a source of nitrogen for growth. This may range from gaseous nitrogen through simple inorganic salts (ammonium, nitrite, or nitrate) to complex organic compounds. A number of differential tests are based on the ability to metabolise various amino acids.

Inorganic ions are also required, the most important ones being $Mg^{2+}$, $Fe^{2+/3+}$, $K^+$, $Na^+$, $Cl^-$, and $PO_4^{3-}$.

Some species require the presence of vitamins and certain amino acids for growth. These are chemicals which the cell is unable to synthesise for itself and which must be supplied in the medium.

The availability of water is of great importance to micro-organisms, and drying products, especially foods, is an important method of preservation. Water activity ($A_w$) is calculated from the equation

$$A_w = \frac{\text{Vapour pressure of the solution}}{\text{Vapour pressure of water at the same temperature}}$$

The higher the percentage of compound in solution the lower the $A_w$. Most micro-organisms grow best in dilute solutions, and as the concentration increases the growth rate falls until a limiting $A_w$ value is reached. This varies from organism to organism, approximate values being Gram-negative rods 0.95, micrococci 0.90, and most yeasts 0.88. Some osmotolerant spoilage yeasts will grow down to $A_w$ values of 0.75, and a few filamentous xerophilic fungi will grow at values of 0.7.

The temperature of growth influences the growth rate considerably, and organisms are divided into three groups, psychrophilic, growing at temperatures of less than 10 °C, mesophilic, growing at 10–45 °C and thermophilic, growing above 45 °C, although the temperature limits quoted are arbitrary.

Most bacteria grow best at pH values in the range 6.0–8.5, although some important species will grow at pH values down to 1.0–2.0. Many fungi will grow at pH values of 5.0–8.0. An important aspect of the medium is the buffering capacity. Many organisms produce considerable quantities of acid during metabolism, and if the medium is not buffered adequately, growth will rapidly cease due to the fall in pH.

Many of the criteria mentioned above may be used when designing media.

Basal medium may be elaborated in various ways depending on the organisms under examination. The most obvious example is the addition of a gelling agent such as agar, an inert marine polysaccharide, to solidify the medium. This is not essential for the growth of the organism, but is added solely for the convenience of the laboratory worker.

Complex media, which are chemically undefined, may be used for the growth of more demanding micro-organisms. Media may be prepared from hydrolysates of meat or yeast. Blood or serum is frequently added to a medium to allow the growth of pathogenic organisms.

Selective media may be prepared by adding compounds such as antibiotics or bile salts, which allow the growth of certain species and suppress the growth of other species. Wide use is made of this type of medium in the identification of unknown species.

Differential media contain chemicals or additives which allow similar micro-organisms to be differentiated from each other. A typical example would be the addition of blood from various species to nutrient agar, to distinguish different types of streptococci.

Media for the assay of vitamins or amino acids and media for counting are frequently defined by a number of bodies such as the Pharmaceutical Society or the Water Boards. It is necessary to comply with these requirements if, for example, produce is to be sold with the BP label.

Many of these media are commercially available, and it is usually only necessary to prepare it in house if there is some unusual requirement.

The growth of bacteria passes through several distinct phases; these include a lag phase, log phase, stationary phase and death phase (Figure 1.7). The lag

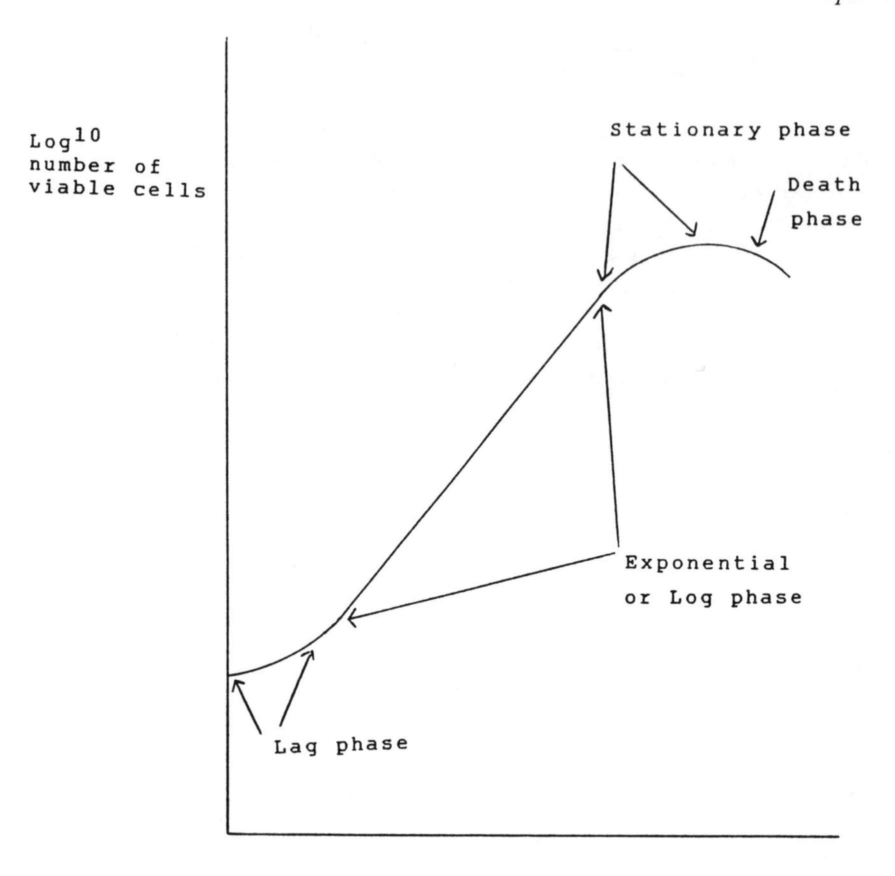


**Figure 1.7** *Plot showing the growth phases of bacteria*

phase is a period of enzyme synthesis during which the cell is becoming accustomed to the medium. The duration is variable, depending on the medium involved, the temperature and the previous history of the organism. This is followed by the log phase during which the organism grows exponentially. Cell division in bacteria is by a process of binary fission, in which each bacterium grows and divides into two daughter cells. The time taken for a cell to divide into two is known as the mean generation time or doubling time. The mean generation time depends on the organism being studied, medium and temperature, and may be as short as twenty minutes for some species growing under optimal conditions. The stationary phase commences when some essential requirement becomes limiting or when some toxic metabolite builds up in the medium. If the cells are held in the stationary phase for a prolonged period, cell lysis commences and cells start to die, releasing their contents which allows limited survival of the remaining cells.

## 4 REFERENCES AND FURTHER READING

1. 'Bergey's Manual of Systematic Bacteriology.' Vol. 1, 'The Gram-negatives of general, medical or industrial importance', ed. N.R. Kreig and J.G. Holt, 1984. Vol. 2, 'The Gram-positives other than Actinomycetes', ed. P.H. Sneath, N.S. Mair, M.E. Sharpe, and J.G. Holt, 1986. Vol. 3, 'The Archaebacteria, Cyanobacteria and remaining Gram-negatives', ed. J.T. Staley, M.P. Bryant, N. Pfennig, and J.G. Holt 1989. Vol. 4 'The Actinomycetes', ed. S.T. Williams, M.E. Sharpe and J.G. Holt, 1989. Williams & Wilkins, Baltimore.
2. 'The Shorter Bergey's Manual of Determinative Bacteriology', 8th Edn. ed. J.G. Holt, Williams & Wilkins, Baltimore, 1977.
3. B.D. Davies, R. Dulbecco, H.N. Eisen and H.S. Ginsberg, 'Microbiology', 4th Edn., Lippincott, Philadelphia, 1990.
4. R.Y. Stanier, J.C. Ingraham, M.L. Wheelis M.L. and P.R. Painter, 'General Microbiology', 5th Edn., MacMillan, Basingstoke and London, 1987.

CHAPTER 2

# Biodeterioration

Microbial spoilage (biodeterioration) covers a wide range of produce and is therefore a very difficult area to treat in one chapter. Most people think of microbial spoilage in terms of food spoilage, but numerous other commodities may also be spoilt.

## 1 ANALYTICAL SAMPLES

Micro-organisms require a source of carbon and nitrogen for growth and any sample containing carbon or nitrogen is potentially subject to biodeterioration. It is important to realise that once the micro-organisms start to grow, they may utilise any carbon and/or nitrogen source present, and thus alter the chemical profile of the sample. It is therefore essential that when a sample potentially capable of biodeterioration is collected for analysis, good microbiological practice should be followed. Good practice consists of collecting the sample in a sterile container, adding a preservative if appropriate, freezing or cooling the sample, and carrying out a chemical analysis as rapidly as possible.

It is important to realise that all compounds formed biochemically can be broken down biochemically, and that micro-organisms are also capable of degrading xenobiotics (unnatural synthetic compounds). It should also be realised that many microbial communities are much more effective at degrading chemicals than the sum activities of their component species.

The rate of deterioration of any product depends on a number of factors, the important ones being the chemical structure, physical characteristics, the type of micro-organisms present, the number of micro-organisms present, temperature, pH, and the water content ($A_w$).

## 2 COSMETICS, TOILETRIES AND PHARMACEUTICALS

Medical or pharmaceutical products are spoiled if they contain pathogens or opportunistic pathogens, if they contain toxic microbial metabolites, or if microbial growth has caused physical or chemical deterioration of the product. It should be realised that an organism that is harmless to a person in good health

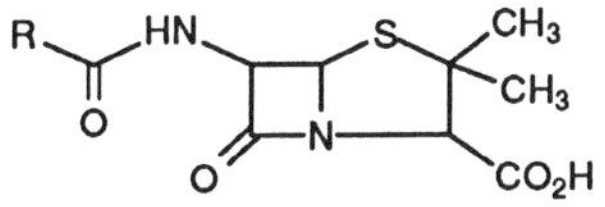

**Figure 2.1** *Structure of penicillin*

may present a very serious problem to someone whose health is compromised, especially if the material is to be administered intravenously.

Specific problems that may arise with medical preparations are the presence of *Pseudomonas* spp. in eye drops, and the presence of *Salmonella* spp. in pharmaceutical preparations which are taken orally, and which are derived from some natural source.

The main problems from microbial metabolites found in medicines are caused by pyrogens, which are lipopolysaccharides (usually). These are derived from Gram-negative bacteria, and can induce acute shock. Other compounds are the mycotoxins formed by fungi, which may be found in medicines derived from natural products. These are considered in more detail in the chapter on food.

Numerous chemicals which will support microbial growth are found in pharmaceuticals and cosmetics. These include sugars used as sweetening agents, polymers such as starch used as thickening agents and for use in suspensions, and oils and fats used in the preparation of creams and emulsions. Growth of microorganisms on many of these products causes the formation of foul smelling odours, and the separation of emulsions into their component parts.

The degradation of creams depends on the presence of water, and a point which is frequently not appreciated is that as one of the major products of microbial metabolism is water, the degradation process becomes autocatalytic. Drugs and antibiotics may also be destroyed by microbial action; for example, penicillin (Figure 2.1) will be destroyed by any β-lactamase producing organism. Many preservatives, disinfectants and surface active agents are metabolised by bacteria, although usually at concentrations below the normal 'in use' concentration. There have, however, been a number of reports of the degradation of simple preservatives such as benzoic acid at 'in use' concentrations.

## 3 FOOD

The biodegradation of food is probably the most common aspect of spoilage. Food spoilage can be defined as any organoleptic change which the customer considers undesirable. This can include visual, tactile, flavour, or olfactory changes.

Only post-harvest decay or spoilage will be considered. Food spoilage organisms are frequently grouped according to the type of activity; for example, proteolytic, lipolytic or saccharolytic, or by growth requirements such as thermophilic or halophilic.

Most problems are the result of poor storage conditions caused by inadequate refrigeration, fumigation, pest control, drying, or ingress of water.

Post-harvest losses are significantly greater in tropical and sub-tropical countries. Estimates relating to Central Africa where temperature and humidity are high, and transport and storage facilities are poor, suggest losses ranging from 40% for avocadoes to 70% for maize and up to 90% for sweet potatoes and soft fruits. Even in the USA, losses of crops such as strawberries have been estimated at up to 25% between harvest and market.

Food spoilage is frequently caused by fungi rather than bacteria. Fungi have been implicated in one disease, ergotism, for over a century, but for a long time the growth of most fungi on food was regarded as merely a nuisance, which detracted from the appearance of the products. In severe cases, produce might be considered unfit for human consumption and used as animal feed. This attitude changed markedly with the discovery of the aflatoxins and other mycotoxins, and it is now therefore necessary to consider fungi not only as spoilage organisms, but also as serious potential health hazards. The use of damaged material for animal feed was seriously re-evaluated, following the destruction of many poultry flocks fed with mouldy cereals or peanuts rejected for human consumption.

The amount of damage caused by fungi is variable and will depend on the type of produce, water content, storage temperature, and the type and number of fungi present.

A number of different forms of deterioration can be recognised. The large volume of work published in the past few years on mycotoxins has meant that the problem of fungal spoilage has been heavily biased in this direction. However, other types of fungal damage can also be recognised and these will be considered first.

The production of lipases by fungi causes the lipolysis of triglycerides producing a rise in the free fatty acid content of seed oils. High free fatty acid values in seed oils cause accelerated auto-oxidation leading to rancidity and browning. This causes increased refining costs, especially those associated with bleaching, and also causes increased losses during processing. A rough calculation is that the oil lost during refining is 1.5 × the free fatty acid value.

The financial penalties caused by these fungal reactions may be high. Many purchase and sale contracts stipulate an oil and fatty acid content. For each 1% above or below these stipulated figures, a bonus or penalty is added or deducted. The free fatty acid changes may be considerable as work in our laboratories has shown that whilst the free fatty acid value of good quality palm kernels is 3.6%, the free fatty acid value of kernels heavily infected with fungi is 29%.

A rise in free fatty acids may also cause changes in the flavour components. Increased β-oxidation of the free fatty acids leads to β-keto acids which are decarboxylated to methyl ketones. There are also increases in unsaturated aldehydes, especially 2-enals and 2,4-dienals, probably formed by oxidation of unsaturated fatty acids. This marked increase in carbonyls and unsaturated aldehydes may cause a significant alteration in the flavour, as these compounds have considerable organoleptic properties.

Rapid fungal growth in seeds may also cause heating which can lead to loss of viability, and in extreme cases may cause charring and spontaneous combustion.

A major cause of fungal spoilage of food is the production of mycotoxins. Over

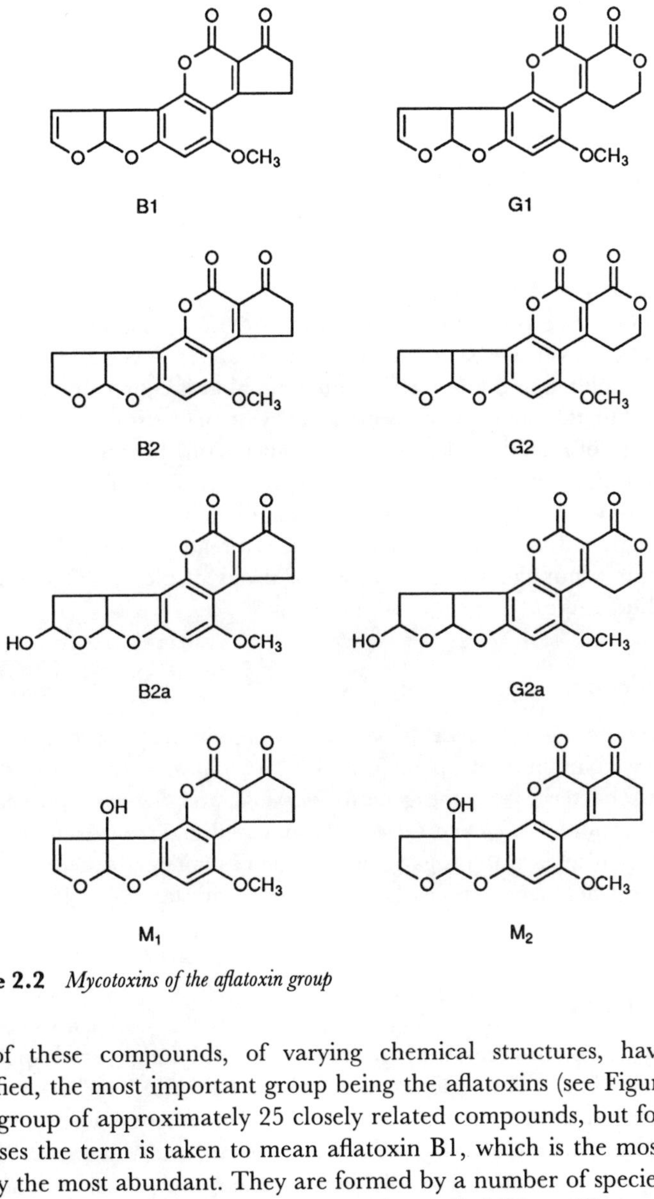

B1 G1

B2 G2

B2a G2a

$M_1$ $M_2$

**Figure 2.2** *Mycotoxins of the aflatoxin group*

200 of these compounds, of varying chemical structures, have now been identified, the most important group being the aflatoxins (see Figure 2.2). These are a group of approximately 25 closely related compounds, but for all practical purposes the term is taken to mean aflatoxin B1, which is the most toxic and is usually the most abundant. They are formed by a number of species of *Aspergillus* and also members of other genera such as *Penicillium.* The major aflatoxin producing species, *Aspergillus flavus,* forms toxins in a wide variety of commodities. Ingestion of aflatoxin contaminated food by farm stock can lead to modified aflatoxins, such as aflatoxin M appearing in milk.

Conditions favouring aflatoxin production are very variable and studies suggest that temperature, substrate, water and a number of other factors may influence production. A recent study has suggested that the epoxides formed from unsaturated free fatty acids may increase aflatoxin production.

**It should be noted that these compounds are extremely toxic and anyone working with them should be aware of the hazards involved.**

Mycotoxins will be considered in more detail in the chapter on microbial counting.

## 4 SPECIFIC FOOD PROBLEMS

### 4.1 Cereals

The usual problem is fungal damage due to inadequate drying. The most common genera causing problems are *Fusarium*, *Aspergillus*, and *Penicillium* in the fungi, and *Pseudomonas* in the bacteria. Most species are removed from the outer seed coat during milling and any further contamination is caused by inadequate storage. The best method of control is sufficient drying before storage.

The appearance of moulds on bread is nearly always due to inadequate storage or dirty bread slicing machines. The main culprits are fungi from the genera *Penicillium*, *Aspergillus* and *Rhizopus*. One interesting spoilage organism is *Serratia marcescens*, a bright red bacterium which grows on bread occasionally, and was responsible for the 'miraculous' appearance of drops of blood on consecrated bread in the Middle Ages.

### 4.2 Sugar Products

Raw juice from cane sugar (sucrose) becomes very high in micro-organisms unless treated promptly. Numbers of up to $5 \times 10^{8}\,ml^{-1}$ have been reported. The main species are bacteria from the genera *Leuconostoc* and *Bacillus*, and yeasts of the genera *Saccharomyces*, *Pichia* and *Candida*. Granulated sugar is extremely low in bacteria. Spoilage of sugars and sugar products such as honey is generally limited to osmophilic species such as yeasts capable of withstanding high osmotic pressures.

### 4.3 Alcoholic Drinks

A wide variety of micro-organisms, both bacteria and fungi, are capable of spoiling wine and beer. This spoilage may take the form of 'off' flavours, or oxidation of the ethanol content to acetic acid. The major bacterial genera causing problems are *Lactobacillus*, *Pediococcus*, *Acetobacter*, and *Leuconostoc*. Fungal species causing problems are generally the so-called wild yeasts and include genera such as *Pichia* and *Schizosaccharomyces*.

### 4.4 Fruit and Vegetables

Microbial damage is frequently accentuated by mechanical damage such as bruising or damage caused by insects or birds. Enzymatic action within the fruit may also contribute in several ways; for example, the conversion of starch to sugars during ripening provides a substrate for microbial growth. Physical damage to fruit also allows the entry of oxygen, which activates phenol oxidases causing browning of items such as potatoes and apples.

It is difficult to be specific about microbial damage in fruit and vegetables because there is usually a succession of micro-organisms and each commodity will differ. However, numerous different types of spoilage can be recognised.

(a) Bacterial soft rot is caused by the genus *Erwinia* and forms mushy products.
(b) Grey rot is a fungal rot caused by *Botrytis* on grapes and other fruit. Its influence on grapes depends on the time of year; for example, if grapes are infected late in the season then it is known as noble rot, and the wine prepared from it is a sauterne.
(c) *Rhizopus* rots produce mushy fruits with growths like cotton wool and small black dots.
(d) Blue mould rots are caused by the fungus *Penicillium.*
(e) *Alternaria* rots are frequently found on soft fruits such as strawberries.
(f) Downy mildew appearing as white woolly masses is caused by the fungus *Phytophthora*
(g) Stem end rot is caused by the fungus *Fusarium* attacking the cut ends of fruit stems.
(h) Souring or slimes is caused by saprophytic bacteria growing in piles of wet vegetables.

Figure 2.3 shows palm kernels which have been attacked by a fungus.

**Figure 2.3** *Good (top) and damaged (below) palm kernels*
(Reprinted from Food Chemistry, **18**, R.K. Dart, E.B. Dede, and J. Offem, p. 115, © 1985, with kind permission from Elsevier Science Ltd., The Boulevard, Langford Lane, Kidlington OX5 1GB, UK)

### 4.5 Meat

Meat spoilage is generally classified depending on whether it is aerobic or anaerobic, and whether it is caused by bacteria, yeasts, or moulds.

*4.5.1 Aerobic Spoilage*
This can take several forms:

(a) A surface slime is formed usually by bacteria, especially species such as *Leuconostoc*.
(b) Meat pigments change due to oxidation by *Lactobacillus* and *Leuconostoc* species.
(c) Rancidity in fats and oils is caused by the bacterial genus *Pseudomonas*.
(d) The observance of bacterial phosphorescence, which is unusual.
(e) The bacterial genus *Serratia* causes pigmentation.
(f) Tainted meat is caused by the production of volatile fatty acids, usually formed by *Actinomyces* species.
(g) The growth of yeasts and moulds may cause several problems. The most common are rancidity, production of black spots due to *Cladosporium* and blue/green spots due to *Penicillium*. A number of fungi are able to grow at temperatures well below freezing.

*4.5.2 Anaerobic Spoilage*
Generally two types are found, souring and putrefaction.

(a) Souring is caused by the formation of volatile fatty acids, usually formed by *Lactobacilli* and certain *Clostridium* species.
(b) Putrefaction is the anaerobic decomposition of protein with the formation of compounds such as $H_2S$, ammonia, indole, and mercaptans. It is generally caused by bacteria from the genus *Clostridium*.

## 5 WOOD (LIGNOCELLULOSE)

Untreated soft woods are rapidly attacked by micro-organisms, especially fungi. Bacterial involvement is usually a secondary process. Hard woods are generally considerably more resistant to biodegradation. The majority of fungi attacking wood are in the groups Ascomycetes or Basidiomycetes, and several types of wood rot can be recognised on the basis of the wood component attacked.

(a) *White Rot*. This is caused primarily by Basidiomycetes and more or less total degradation of both the lignin and cellulose fractions of the wood takes place.
(b) *Brown Rot*. These organisms degrade cellulose without attacking lignin to any great extent.
(c) *Soft Rot*. This is generally caused by members of the Ascomycetes or Fungi Imperfecti. These organisms degrade cellulose in wet conditions and will attack wood under conditions of oxygen limitation.

There are also a number of organisms which do not fall into one of the above groups, these include the Actinomycetes which are bacteria, not fungi.

## 6 ANIMAL PRODUCTS (NON FOOD)

This group of products covers wool, hides and some types of glue. All these materials are basically protein and will be attacked by fungi or bacteria possessing proteolytic enzymes.

Glues of animal origin are very prone to biodegradation and will fail rapidly in hot damp conditions.

Freshly cut wool is high in lipids, and is subject to rapid decay unless it is degreased quickly. Generally, once it has been prepared adequately, wool does not suffer severe deterioration problems.

Fresh hides are also very high in lipids and deteriorate rapidly if stored in heaps. Finished leathers will deteriorate rapidly in hot damp climates.

## 7 PAINTS

Paints frequently suffer from microbial damage. There are basically two types of paint, solvent thinned and water thinned. Water thinned paints contain cellulose derivatives as viscosity stabilisers, and attack by cellulolytic fungi will cause severe damage. Solvent thinned paints are generally microbiologically stable in the tin. Dried films of both types of paint are susceptible to attack by bacteria, fungi and algae. Solvent based paints are attacked by micro-organisms producing lipase enzymes, which degrade the linseed oil present in these paints to form free fatty acids. Once formed, the free fatty acids can degrade further.

Control is usually achieved by the addition of biocides, although these must be chosen carefully to avoid destroying the properties of the emulsion.

## 8 RUBBER AND PLASTICS

Natural rubber is a polymer of isoprene units. As a naturally occurring polymer, it is subject to microbial attack especially in hot humid conditions. The microbial degradation may be accentuated by chemical oxidation.

Plastics vary in their susceptibility to micro-organisms. Plastics based on polyesters are attacked fairly rapidly, whilst the polyether based plastics are much more resistant. However, the situation is a very complex one, as many of the plasticisers and vulcanisers used in the production of plastics are able to support microbial growth. Once these have been removed, the plastic frequently becomes brittle irrespective of type.

## 9 FUEL AND LUBRICANTS

These are basically hydrocarbons and are not susceptible to microbial attack in the absence of water. When water is present, considerable growth at the water/fuel interface may occur. If the growth becomes detached it may move into the filter systems causing a blockage with disastrous results in the case of aircraft. The worst culprit is a fungus of the genus *Cladosporium*. This situation may arise in any aircraft which is grounded for long periods of time with fuel in its tanks, a situation frequently found in military aircraft held in a state of combat readiness. It may also arise in supersonic aircraft during flight, when the heat generated on wing tanks is high enough to allow the growth of thermophilic *Nocardia* and *Aspergillus* species. It does not generally arise in subsonic aircraft, as the temperature during flight does not rise sufficiently to allow microbial growth, and there is a rapid turnover of the fuel. This problem may also occur in fuel storage tanks.

In addition to the problem of filter blockage, the acidic products of metabolism may also cause corrosion of metal tanks and pipes. Biocides may be added to the fuel, but care has to be taken to avoid compounds which might reduce the performance of the fuel, or lead to corrosion of the fuel tanks or engines.

Similar problems arise with many metal working fluids. These are sprayed onto the cutting tools, fall into an open collecting vessel, and are recirculated. They are frequently contaminated with organic detritus such as food crumbs, and mice or rat faeces or urine. They may be topped up periodically with fresh oil and maintained at temperatures of 20–30 °C leading to rapid deterioration. The effect of bacterial metabolism is to break the emulsion into organic and aqueous layers, the oil is converted into fatty acids, the pH falls the emulsion becomes useless and may cause corrosion spots on the finished item.

## 10 METALS

Micro-organisms can corrode susceptible metals by the metabolic production of chemicals which can attack the metal. For example the oxidation of sulfur and sulfur containing compounds produces sulfuric acid, generating pH values as low as 1.0. Sulfate reducing bacteria, growing under anaerobic conditions, produce sulfides which cause rapid corrosion of iron based products.

## 11 STONEWORK

Penetration of stonework by fungi and lichens (combinations of algae and fungi) can lead to minute cracks which allow the penetration of water. Subsequent freezing can lead to larger cracks allowing further penetration and rapid deterioration.

Considerable damage is also caused to concrete and marble by sulfur oxidising bacteria capable of producing acids.

## 12 FURTHER READING

'Industrial Microbiological Testing', Society for Applied Bacteriology, Vol. 23, ed. J.W. Hopton and E.C. Hill, Blackwell Scientific Publications, Oxford, London, and Edinburgh, 1987.

CHAPTER 3

# Equipment and Methods

## 1 THE LABORATORY

The number of laboratories used will depend on the space and funds available, and also on the estimated number of samples. There should be at least one laboratory available for the counting and identification of bacterial samples, with a separate media kitchen or clean area for the preparation and sterilisation of media. A separate area should also be available as a dirty area for the disposal of used materials.[1] It has to be accepted however that it is not an ideal world, and frequently a new laboratory has to be fitted into available space rather than designed from the beginning on first principles.

Benches in the laboratories should be impervious to liquids and resistant to solvents, such as ethanol, and chemical disinfectants, such as hycolin. The ceilings and walls should be capable of being cleaned easily, and laboratories should be provided with water, gas, and electricity.

There should also be a separate area available for staff to wash their hands before leaving the laboratory. Taps in this area should preferably be worked by the elbows.

The most important aspect of laboratory design is safety and this should be of paramount importance. Safety in the laboratory is discussed by Darlow[2] and laboratory organisation and management by Duguid.[3]

## 2 EQUIPMENT

The equipment required will obviously depend on the number of samples expected, but it should be remembered that the number of samples will probably exceed expectations over a period of time.

A minimum level of equipment that needs to be considered is as follows:

(a) Binocular microscope, slides, and coverslips.
(b) Autoclave. Considerable attention should be paid to size and turn round time. It is advisable to obtain one that is larger than initially required to allow for an increasing number of samples over a period of time. As large autoclaves are heavy, the floor loading capacity is an important factor. It is

also useful to obtain a small autoclave with a short turn round time to allow for quick sterilisation in emergencies.

(c) Incubators. Four incubators should be regarded as the minimum to allow operations at 22/25, 30, 37 and 55 °C. Larger incubators will be needed at 22, 30, and 37 °C rather than at 55 °C.

(d) Water baths. Three of these are required to operate at 30, 37, and 44 °C. A fourth one operating at 55 °C for keeping agar molten is also required, although this may be replaced by an oven at the same temperature. A boiling water bath for re-melting sterile agar is also useful and is considerably quicker than using an autoclave for the same purpose.

(e) A hot air oven with a range up to 200 °C is required for sterilising glassware.

(f) A refrigerator is needed for the storage of heat labile materials and also for the overnight storage of samples which cannot be analysed immediately.

(g) A centrifuge may be useful, but the requirement for this will vary from laboratory to laboratory depending on the type of samples being handled.

(h) A glove box or laminar flow hood is also useful for carrying out work in a non-sterile environment. These vary widely in price and complexity and the principles involved are discussed by McDade.[4]

(i) Membrane filtration equipment.

(j) Anaerobic jars. The necessity for these can frequently be removed by using the appropriate media.

(k) Automatic media dispensers. These are not essential, but they save considerable preparation time and are cost effective if large quantities of media are being handled.

(l) Blenders/stomachers for the homogenisation of samples.

(m) Colony counting equipment. Both manual and automated devices are available.

(n) Automatic pipettes (in the range 10 μml–10 ml) with appropriate disposable tips. A range of pipette sizes should be available for each worker.

(o) Staining racks and trays for each worker.

(p) Bunsen burner and tripod for each worker.

(q) Inoculating wires/loops, forceps, scissors, and waterproof marker pen for each worker. Sterile disposable loops are available.

(r) Two disinfectant jars for each working space, one for discarded slides and the other for pipette tips.

(s) Racks and baskets for test tubes and screw capped bottles.

(t) Sterile disposable petri dishes.

(u) An adequate supply of disposable gloves should be available.

(v) There should be two Howie style laboratory coats available for each worker to allow for laundering.

The following glassware should be available.

(a) Petri dishes. Whilst a few glass ones are needed where solvents are in use, the problems of cleaning and resterilising glass petri dishes make it much more cost effective to use ready sterilised, disposable, plastic petri dishes.

(b) Test tubes, boiling tubes and Durham tubes.
(c) Screw capped 'Universal' bottles.
(d) Disposable Pasteur pipettes.
(e) Storage bottles of various sizes which are autoclave proof.
(f) A selection of beakers, conical flasks and measuring cylinders of various sizes.

Whilst this is not a comprehensive list of equipment and glassware, it is adequate to cover most situations.

## 3 PERSONAL HYGIENE

Eating, drinking and smoking should be totally banned in the laboratory. Staff should also undergo an induction course to enable them to realise the fact that micro-organisms are ubiquitous, and if suitable precautions are not taken when handling samples then contamination is inevitable.

## 4 METHODS

Micro-organisms in nature exist in mixed populations and one of the first essentials is to isolate or obtain a pure (axenic) culture. Pure cultures may either be isolated or if reference material is required they can be obtained from various culture collections. Isolation of a pure culture is only the first step; it is also necessary to culture and maintain the organism in the pure state in which it is protected from contamination by other micro-organisms.

## 5 PREPARATION OF PLATES

Petri dishes containing solid media may be prepared as follows. Ten ml of a medium such as nutrient agar is placed in a universal screw capped bottle and sterilised by autoclaving. Once removed from the autoclave the plates may be poured immediately, or the bottles of agar may be placed in an oven at 50–55 °C to keep them liquid. Agar has the unusual property in that it does not melt until it is heated to 100 °C, but then remains liquid until the temperature falls to about 50 °C. When pouring plates, the outside of the universal bottle should be wiped to remove any water, the cap removed, and the neck of the bottle passed through a bunsen flame. The lid of the petri dish should be removed and held over the lower part to prevent dust falling into it and the agar poured into the small part of the petri dish. The lid is then replaced and the petri dish should be allowed to stand on a level spot until the agar has set. The plates may be opened and placed in an inverted position in a 37 °C incubator if they need to be dried.

The quantity of agar normally placed in a petri dish is 10 ml but more may be used if required. This is usually done when a long incubation period is required as plates containing only 10 ml of medium would dry out.

## 6 ISOLATION OF PURE CULTURES

Staff should always work beside a lighted bunsen burner. This is the most convenient method of sterilising materials and also provides an up-draught which stops dust falling onto media and materials.

Streak plates are the most useful method of isolating organisms. A suitable nutrient medium containing agar is poured into a sterile petri dish and allowed to set and dry. A wire inoculating loop is sterilised by passing it through a bunsen flame, or alternatively a sterile plastic disposable loop is used. Wire loops should always be placed into the coolest part of the flame first and then drawn up into the hot portion. If placed directly into the hot region, there is a distinct tendency for material on the loop to spit out of the flame and not be destroyed. The sterile loop is dipped into the sample from which the organism is to be isolated, and is then drawn gently across the surface of the agar several times to cover an arc of about sixty degrees. The loop is resterilised in the bunsen flame or in the case of plastic loops, a fresh one is used. The loop is then drawn through the tail end of the previous streaks across a second arc of sixty degrees. This is repeated a third and fourth time. The effect is to progressively dilute the sample until individual well isolated organisms are obtained (see Figure 3.1). When incubated at the appropriate temperature, these individual organisms will grow to form single colonies which, with care, can be picked off the plate. In theory, these should give a pure culture when grown in the appropriate medium. However, if time allows it is better to streak out the colony onto a second plate. The medium should be a general medium such as nutrient agar which allows the growth of a wide variety of organisms. Selective media should not

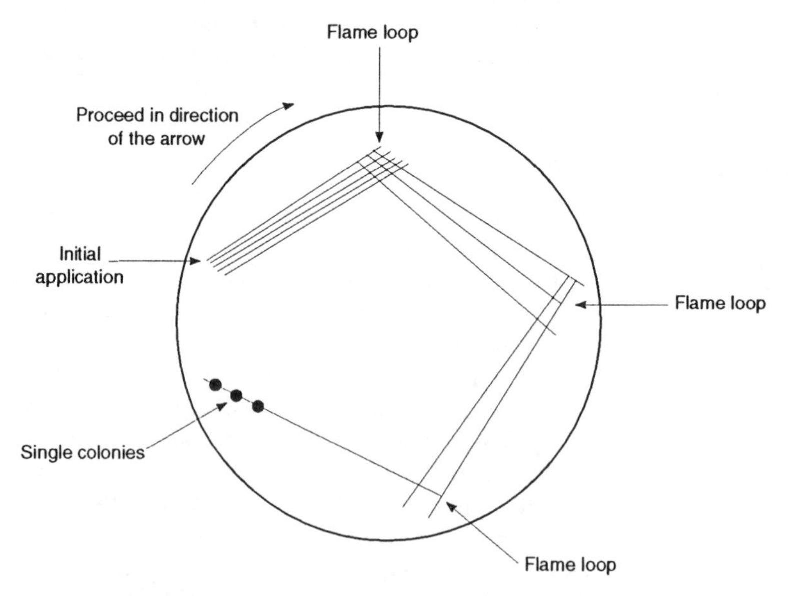


**Figure 3.1** *Inoculation of a petri dish using a wire loop. The wire is flamed at the points indicated resulting in a progressive dilution of the sample until single colonies are obtained. If plastic disposable loops are being used, then the loop should be discarded at the points where a wire loop would be flamed and a fresh plastic loop should be taken*

be used for isolation purposes as they frequently only suppress the growth of contaminating organisms and do not kill them. The contaminating organism may then grow through at a later date.

A second method of isolating micro-organisms used less frequently is serial dilutions. The sample is diluted repeatedly in a sterile diluent until there is theoretically only one organism per tube. This is then grown to give a pure culture although the purity should be checked by plating out as described previously.

In the case of an organism such as *Bacillus*, considerable purification can be achieved by boiling the sample, which would kill all organisms except the resistant spores of the genus *Bacillus*.

## 7 MAINTENANCE OF PURE CULTURES

Once purified, cultures are best maintained on slopes of a nutrient medium such as nutrient agar. These are prepared by placing 10 ml of medium containing agar in a universal screw capped bottle. This is then sterilised and allowed to set at an angle to give 'slopes'. If only a few bottles are involved this is best achieved by laying the necks of the bottles across horizontal test tubes. If a large number of bottles are required, then they are best placed in a rack which is held at an angle until the medium has solidified.

Once the medium has solidified, it should be inoculated as follows. Flame a wire inoculating loop (or use a sterile plastic one) and pick a colony from the streak plate, holding the loop in the right hand (if right handed). Hold a slope in the left hand and remove the screw cap using the little finger of the right hand (it is best to loosen the screw cap slightly before commencing). Do not put the cap down. Pass the neck of the bottle containing the slope through a bunsen flame and draw the loop containing the bacteria lightly across the surface of the agar slope. Flame the neck of the bottle again and replace the screw cap. Flame the inoculating wire. The slope can now be incubated at the appropriate temperature for a sufficient length of time for the organism to grow. Once grown, the slopes can be stored in a refrigerator until needed. It is best for new staff to practise this procedure several times on unimportant material before attempting it on real samples.

Several slopes of each organism should be prepared. One of these should act as the master culture and should not be used unless it is essential. The function of the master culture is to preserve the culture with all its original characteristics, and it should only be used if contamination of the working cultures takes place, or if the strain is lost. The other slopes are the working slopes which are used on a day to day basis. Master slopes can be kept in a refrigerator for a considerable period of time varying from two weeks to six months depending on the organism involved. There should be a regular programme of reinoculating master slopes at the appropriate intervals, and also preparing streak plates to check the purity of the organism.

## 8 LABORATORY STOCKS OF BACTERIA

Micro-organisms are frequently supplied as freeze dried cultures *in vacuo* in sealed ampoules. A list of suppliers is given in Appendix 2. The freeze dried cultures should be recovered into a nutritionally rich liquid medium several days before being required, as organisms may need several successive subcultures before regaining their normal physiological characteristics.

The process of opening ampoules and recovering freeze dried cultures can be difficult. The ampoule should be opened with a glass knife and a few drops of sterile nutrient broth introduced with a Pasteur pipette. The contents should be sucked up and down carefully several times and then placed into a universal bottle containing nutrient broth or some other suitable growth medium for the organism in question. A drop of culture should also be placed on a solid medium such as nutrient agar to check for purity. Once the culture has grown on the agar plate, it should be subcultured onto a plate of selective/diagnostic medium and checked by examining its characteristic cultural morphology and staining reactions. Once the freeze dried cultures have been recovered and their purity checked, master slopes and working slopes should be prepared.

For most species of bacteria, subculturing the master cultures once a month is adequate. Lactobacilli may require subculturing more frequently, whilst yeasts and flagellates such as *Euglena*, can be subcultured every 3–6 months.

## 9 MEDIA

A large number of different media are used in microbiology. Whilst these can be prepared from the individual ingredients, there are a number of companies specialising in the production of media in a powdered form, many of which can be stored indefinitely. It is only necessary to add water to these and once autoclaved they are ready for use.

## 10 GROWTH TECHNIQUES

The techniques of growing micro-organisms in bulk are not usually required in an analytical laboratory. If organisms are required in moderate quantities (1–2 litres) they are best grown in conical flasks. This can be carried out as follows. Two flasks should be prepared, one of about 100 ml capacity and the other of one litre capacity. Liquid medium should be placed in both flasks to a maximum of about 40% capacity. Cotton wool bungs are prepared for the flasks by rolling cotton wool in a square of muslin and tying the opposite corners of the muslin. These bungs are protected during autoclaving by a square of autoclave paper held in place by a rubber band. After autoclaving the autoclave paper is discarded.

Once prepared the medium is sterilised by autoclaving. When the medium has cooled, the small flask is inoculated aseptically with a loopful of organisms taken from a slope. The flask is then incubated at the appropriate temperature for sufficient time to produce cloudiness of the medium. This may be done statically or in a shaking incubator or water bath. The aseptic inoculation of the small flask is carried out in a very similar manner to the inoculation of the slopes. A sterile loop is held in the right hand and the slope in the left hand. The cap of the slope is removed using the little finger of the right hand, the neck of the slope is flamed and a loopful of bacteria is taken. The neck of the slope is flamed again and the cap is replaced. The conical flask is then taken in the left hand and the cotton wool bung is removed with the little finger of the right hand. The neck of the conical flask is flamed, the loopful of organisms is inserted into the liquid medium and swirled around several times. The wire loop is removed, the neck of the flask is flamed and the cotton wool bung is replaced. The loop is then flamed to destroy any residual micro-organisms.

Once the organisms in the small flask have grown adequately, they may be used to inoculate the large flask. The small flask should be removed from the incubator/water bath and any water on the outside should be removed. The small flask is held in the right hand, the bung is removed with the little finger of the left hand and the neck of the flask is flamed. The large flask is then picked up in the left hand, the bung is removed with the little finger of the right hand and the neck of the large flask is flamed. The contents of the small flask are then poured into the large flask, the neck is flamed and the bung is replaced. The bung of the small flask is replaced and the small flask may now be discarded for autoclaving. It is best to practise this technique on flasks containing sterile medium several times before attempting it on flasks containing living cultures. These sterile flasks may then be incubated at 30 °C and if growth appears further practice is needed at the technique.

The large flask is incubated at the appropriate temperature and time. This is best carried out on a shaking incubator of some type as growth for most organisms is heavier when actively aerated. Once sufficient growth has been obtained, the organisms may be harvested either by filtration through a membrane filtration system or by centrifuging. If centrifuging is used, a force of 5–6000 g for 15 minutes is adequate to harvest most bacteria, whilst a force of 2–3000 g for the same time is adequate for yeasts.

Some micro-organisms grow best under anaerobic growth conditions. The requirement varies from organism to organism. Some bacteria, such as *Clostridium*, which are considered anaerobic will grow aerobically if grown on a medium which is sufficiently reducing such as Reinforced Clostridial Medium. Other organisms, however, require the removal of oxygen from the growth medium. This may be achieved by growing cultures in an anaerobic jar. These are jars from which air can be removed physically by means of a vacuum pump, or chemically by means of a catalyst or chemical pack which can be placed in the jar. The manufacturer's instructions should be followed.

## 11 PIPETTING

The use of glass pipettes should be avoided if possible and no mouth operations should be allowed. If it is necessary to use glass pipettes, then these should have a small piece of cotton wool introduced into the mouth end and then be sterilised in cans in an oven. With practice, it is possible to remove the lid of the can, shake out the pipettes, and remove one without touching the others. The lid can then be replaced keeping the other pipettes sterile. These pipettes may then be used with a pipette filler. It requires considerable experience to keep these glass pipettes sterile and avoid dropping contaminated liquid during manipulations. Once used the pipette should be placed into a container of disinfectant.

The introduction of automatic pipettes and disposable tips has made pipetting significantly easier. The pipette can be set to the required volume before commencing and after use the tip is ejected into a container of 2% hycolin or similar disinfectant.

Any contaminated liquid spilled as a result of pipetting or any other operation may be mopped up using a high quality paper towel soaked in 2% hycolin. The paper towel should then be discarded into the autoclave bag kept at the end of each bench (see later).

## 12 AUTOCLAVING AND STERILE PREPARATION

Death in micro-organisms can be defined as the irreversible loss of reproductive ability. As will be seen later, the death of bacteria is exponential, and in practical terms sterility is defined as the probability that the material treated should not contain a survivor in an infinitely large sample. This is usually defined in practice as the probability of less than one bacteria per million units (1 in $10^{-6}$).

Thermally stable material such as glassware or metal instruments may be sterilised by heating them in an oven at 185 °C for two hours. The material is wrapped in autoclave paper prior to heating, and after removal remains sterile until the wrapping paper is removed. Steam treatment in an autoclave is normally used for the sterilisation of aqueous material. The autoclave uses steam at a pressure greater than atmospheric and laboratory systems normally operate at 15 lbs $in^{-2}$ which corresponds to a temperature of 121 °C. This makes the assumption that the atmosphere inside the autoclave is composed only of steam and therefore it is necessary to expel all the air before the sterilisation process commences.

The time taken to sterilise materials depends on the load, but typically loads made up of relatively small volumes would be treated adequately within 20 minutes. Loads containing large volumes take longer and suitable times can be determined by the use of thermal probes or spore strips (see page 86).

The actual operational sequence of autoclaves varies considerably from manufacturer to manufacturer and the appropriate instructions should be followed.

Material which is thermally unstable cannot be sterilised by autoclaving, and solutions should be sterilised by membrane filtration through a membrane with a pore size of less than 0.45 μm. Membrane filtration is dealt with in more detail in the chapter on water testing. It should be noted that membrane filters will only remove bacteria and do not necessarily remove viruses.

Thermally unstable material may be added to autoclaved material aseptically once it has cooled.

## 13 DISPOSAL OF CONTAMINATED MATERIAL

All sharps such as hypodermic needles should be placed in a special container for disposal by a specialist unit. Hypodermic needles can be mutilated by the use of a special guillotine which removes the needle from the barrel.

All benches should have a disposable plastic autoclave bag at the end into which material such as contaminated plates can be placed for autoclaving. When full these should be autoclaved according to the recommendations of Howie.[5] The local authority should also be consulted before disposing of any materials even though they have been adequately sterilised.

## REFERENCES

1. E.C. Elliot and D.L. Georgala, 'Sources, Handling, and Storage of Media and Equipment', in 'Methods in Microbiology', Vol. 1, ed. J.R. Norris and D.W. Ribbons, Academic Press, London and New York, 1969.
2. H.M. Darlow, 'Safety in the Microbiological Laboratory', reference 1, pp. 169–204.
3. J.P. Duguid, 'Organisation of the Clinical Bacteriology Laboratory', in 'Mackie and McCarteney, Practical Medical Microbiology', 13th Edn., ed. J.G. Collee., J.P. Duguid, A.G. Fraser and B.P. Marmion, Churchchill Livingstone, Edinburgh, London, and New York, 1989.
4. J.J. McDade, 'Principles and Applications of Laminar-flow Devices', in 'Methods in Microbiology', Vol. 1, ed. J.R. Norris and D.W. Ribbons, Academic Press, London and New York, 1969.
5. J.H. Howie, 'Code of Practice for the Prevention of Infection in Clinical Laboratories', DHSS Memorandum, HMSO, London, 1978.

CHAPTER 4

# Microscopy and Staining

## 1 MICROSCOPY

The microscope is probably the most essential piece of equipment in microbiology, and its performance can be improved considerably by an understanding of its construction and the associated theory.[1] There are a wide range of microscopes available and the manufacturer's instructions for use should be followed. There is generally a fairly close correlation between performance and price. The microscope is constructed of a number of parts, and consists basically of a system of lenses arranged over a stage, on which the specimen is placed, with a light source illuminating the stage.

There are microscopes with only one eye piece (monocular microscopes), which are useful for demonstrating to classes, but binocular microscopes with two eye pieces are more suitable for laboratories where a great deal of microscopic work is carried out, as they reduce eye strain. The eye piece is usually a lens giving a ten-fold magnification. In the case of a binocular microscope, the two eye piece lenses can be adjusted by the operator for the optimum distance between his or her eyes.

There is then a second lens called the objective, and frequently three or four objective lenses are mounted on a revolving turret. The strength of these lenses varies, but typical ones would give magnifications of 10 ×, 40 × and 100 ×. The values are normally engraved on the casing of the objective. The most useful objectives in microbial work are the 40 × and 100 ×.

The specimen is placed on a stage which can be moved up and down to focus it. The specimen can also be moved in two directions horizontally, so that different fields can be examined. There is a hole in the stage through which light, generated by a built-in lamp situated in the base of the microscope, can pass. The beam of light passes through a condenser (located underneath the stage), which is used to focus the light on the specimen. The amount of light reaching the specimen can be adjusted by means of an iris located in the condenser.

The objective of magnifying a specimen is to produce a clear, well defined image of a specimen too small to be seen with the naked eye. The magnification obtained is the number of times the length of the specimen is multiplied. At a practical level, the easiest way of obtaining the magnification, without considering microscope theory, is to multiply the strength of the eye piece lens by the strength of the objective lens, giving magnifications of 100, 400 and 1000 × for most systems. A magnification of 1000 × is about the largest useful magnification that can be obtained using a light microscope. Above this level, the sharpness of the image is lost and the specimen becomes increasingly blurred.

The 10 × and 40 × objectives are 'dry' objectives, whilst the 100 × is an oil immersion lens. When a dry objective is used, air is present between the specimen and the objective, and light passing through the glass into the air is bent away from the objective and does not enter the microscope. This is not important with the 10 × and 40 × objectives, as sufficient light enters the objective for the specimen to be seen. However, when a 100 × objective is used, this is not the case and immersion oil must be used. If immersion oil (with the same refractive index as the glass slide) is placed on the slide, and the 100 × objective is placed in the oil, then light rays passing through the slide are not bent and enter the objective. The oil immersion lens is used widely in microscopy and the oil should always be removed immediately after use with a very soft lens tissue. Dried lens oil may be removed with a lens tissue moistened with xylol, but no other solvent should be used as it damages the glue holding the lens in place.

There are a number of terms associated with microscopes which are used to define their performance. These include the *numerical aperture*, which at its simplest is the ratio of the diameter of the objective lens to its focal length. This defines the ability of the objective to gather light.

The *limit of resolution* of the microscope is the minimum distance needed between two cells on the specimen slide that allows them to be seen as separate objects and not fused. This distance can be related mathematically to the wavelength of the light used and the numerical aperture of the objective lens. The best optical systems available will give a limit of resolution approaching 0.25 μm, which is adequate for the study of bacteria but not viruses.

The *definition* of the microscope relates to the clearness of the specimen. This requires the removal of optical imperfections found in simple lenses. These include 'spherical aberration' which is caused by the fact that light passing through the centre of the lens travels a different distance to that passing through the edge of the lens. This produces a focal point which is not sharp and causes the image to be blurred. 'Chromatic aberration' is caused by the separation of white light passing through the lens into its component colours, each of which has a different wavelength, and is therefore bent by a different value when it leaves the lens. The result is that the focal point is not sharp, and any image is blurred and edged by the colours of the spectrum. These problems can be solved by the use of complex lens systems known as 'achromatic objectives', which are, however, very expensive.

The microscope should be set up for use as follows. The light source is switched on and the iris in the condenser is closed to its minimum. The height of the condenser is adjusted until the image of the iris appears in the microscope field; this image is adjusted until it is central. The iris is then opened fully and the condenser is raised until maximal brightness is obtained. Whilst filters are not essential, some microscopes come supplied with a pale green filter which reduces glare and eye strain.

Once the microscope has been set up as above, the slide containing the specimen should be placed on the stage and clipped in firmly. The slide should be viewed using a low power dry objective and a suitable field for further examination should be found by moving the slide in a horizontal plane.

Once a suitable field has been found, the low power objective should be moved out of line, a drop of immersion oil should be placed on the specimen, and the oil immersion objective should be swung into line. In a good microscope the objectives should be confocal, that is, if the low power objective is in focus then the high power and oil immersion objectives should also be more or less in focus, when the turret is turned to bring them into line. Slight adjustments can be made to the focus using the fine adjustment. Care should be taken to avoid winding the objective through the slide holding the specimen, although most objectives are spring loaded to stop this occurring.

Novices may have considerable difficulties in focusing with the oil immersion lens. The most common faults are placing the slide on the stage upside down, looking at a field with insufficient material on it to be visible, and using immersion oil that is so old it has become sticky, causing the slide to stick to the objective when attempts are made to focus. With regard to the first fault, the thickness of the slide is such that if it is upside down the objective cannot be brought close enough to the specimen to focus. For the same reason, it is impossible to use a specimen covered with a cover slip when using the oil immersion objective. With regard to the second fault, the best procedure is to return to the low power and find a suitable field. If this cannot be done, then a fresh specimen should be prepared. The solution to the third problem is to clean the slide and objective and use fresh oil.

A transient problem that may occur is a dark shadow passing across the field. This may be due to an air bubble in the oil. The best method of dealing with this is to break the contact between oil and objective, and refocus.

The methods described above are used for studying dry specimens which have been stained. They are not adequate for studying wet specimens of living cells as the contrast with the background is poor. This type of specimen is best studied using dark-field illumination or phase-contrast microscopy.

Dark-field illumination requires a special condenser which produces light rays that are too oblique to enter the objective, unless they hit some specimen material which scatters them. These scattered rays enter the objective and the specimen is illuminated against a dark background.

Phase-contrast microscopy also requires a special optical system. The speed of light passing through objects varies depending on their refractive index. Light passing through a specimen of high refractive index will be retarded compared with light passing through the suspending medium, causing a phase change. This can be detected and enhanced, thus producing a marked contrast between the cells and their surroundings.

When using either of these systems it is best to refer to the manufacturer's instructions, as the techniques of obtaining the best results differ from microscope to microscope.

## 2 STAINING

The refractive index of micro-organisms differs only slightly from the medium in which they are growing, and it is therefore necessary to stain them or use special methods such as dark-field illumination to see them.

A number of stains are used in microbiology, but a common factor before staining is the preparation of a film of bacteria on a slide. The slide should be grease-free. This can be achieved by immersing the slides prior to use in a beaker of alcohol, removing the slide with forceps when required, and passing it through a bunsen flame. Before preparing a smear and staining, it is advisable to label the slide with a waterproof grease pen.

A smear is prepared by taking a loopful of broth or liquid culture containing the bacteria and spreading it thinly on the slide. The most common mistake is to take too much material and prepare too thick a smear. The slide is then held in forceps and passed quickly through a bunsen flame several times to heat fix the organisms. The liquid on the slide should be dried slowly and not allowed to boil. In the case of bacteria taken from a solid medium, a loopful of distilled water should be placed on the slide initially, then a small quantity of bacteria should be removed from the solid surface and emulsified into the water. This is then heat fixed as above. With practice, it is possible to get two or three smears on a single slide. The technique of placing three smears on a single slide is useful, as it is then possible to place the unknown in the centre of the slide, with material known to give a positive result on one side and material giving a negative result on the other.

Once the slide has been prepared, it is best stained on a staining rack over a tray. The slide should be flooded with sufficient stain to avoid drying out during the period of application. Once the staining process is finished, excess liquid may be removed with blotting paper. A wide variety of stains are available, and only the most common or useful will be considered.[2] The structures of some of these are shown in Figure 4.1.

### 2.1 Methylene Blue

This is one of the so-called simple stains, and the version usually used requires Loeffler's methylene blue. This is made by adding a saturated solution of methylene blue in ethanol (300 ml) to one litre of 0.01% aqueous KOH. Smears are stained for three minutes and then washed with water.

### 2.2 Gram Stain

This is the most important stain in bacteriology and is so central to identification that it should be practised until the operator is fully competent. A number of different variations are found,[2] and the laboratory should standardise on one method.

**It is essential that young cultures (24 hrs) are used as old cultures may give false negative results.**

Prepare a smear as described above.

(a) Cover the slide with an ammonium oxalate, crystal violet preparation and leave for one minute. This is prepared by dissolving 20 g of crystal violet in 200 ml of methanol, and adding 800 ml of 1% ammonium oxalate.
(b) Rinse well with water and blot.
(c) Add an iodine preparation and leave for one minute. This is prepared by adding 1 g of iodine and 2 g of potassium iodide to 300 ml of distilled water.
(d) Rinse well with water and blot.
(e) Decolourise with 95% ethanol for approximately 15 seconds, rinse with water and blot.
(f) Stain with 0.5% safranin for approximately 15 seconds, wash with water, blot dry, and examine with an oil immersion objective.

Gram-positive cells will be seen as purple, whilst Gram-negative cells will be seen as pink. A few organisms are Gram-indeterminate, and the best way of examining these is to take a slide and divide it into three portions. A smear of the culture under examination is placed in the middle, whilst a smear of a known

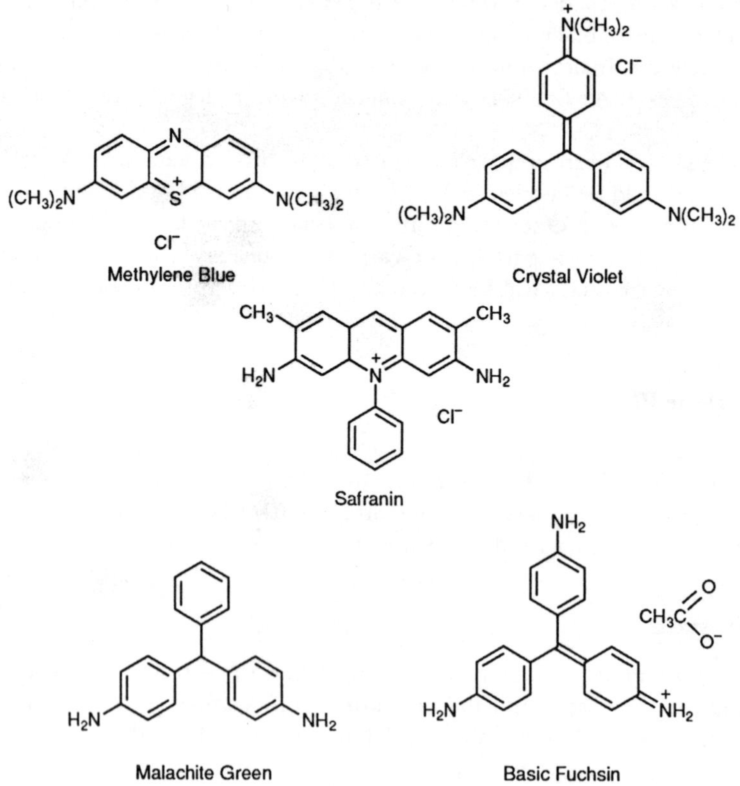

**Figure 4.1** *Some of the wide variety of stains available*

Gram-positive is placed on one side of it, and a smear of a known Gram-negative is placed on the other side. The entire slide is then stained and the central smear can be examined and compared with known organisms treated in exactly the same way.

## 2.3 Spore Stain

Some Gram-positive bacteria produce endospores which are highly resistant to heating and a variety of chemicals. These spores are found in the genera *Bacillus* and *Clostridium.* The spores are difficult to stain using normal techniques, but it is possible to drive the stain into them using heat. Once the stain has been taken up by the spores, it is extremely difficult to decolourise them, although the vegetative portion of the cell can be decolourised easily. Several versions of the spore stain are used.

It is advisable to use old cultures grown on nutrient agar, as spore production is a survival mechanism found in ageing cultures. A smear should be made and fixed in the normal manner. The slide should be placed on the rim of a beaker half full of boiling water, with the bacterial film uppermost, and then the following procedure used.

(a) Flood the slide with 5% aqueous malachite green and leave for 2–3 minutes. Do not allow the stain to dry; if the stain starts to dry, add more.
(b) Remove the slide from the heat and wash well with cold water. Blot dry.
(c) Stain with 0.5% safranin for 30 seconds.
(d) Rinse quickly in cold water, blot and examine under an oil immersion lens.

The spores will stain green and vegetative portions of the cell will stain red/pink. If a very old culture was used, it may be difficult to find vegetative material. The size and position of the spores should be noted, as this may be of diagnostic value, especially when examining members of the genus *Clostridium.* They may be polar or central, and may be the same diameter as the cell or larger than it, causing it to bulge.

## 2.4 Acid-fast Stain

This stain is used to distinguish Gram-positive rods of the genus *Mycobacterium* and, to a lesser extent, *Nocardia.* It was used originally to distinguish tubercle bacteria. Organisms in these groups contain waxy materials in their cell walls which makes them difficult to stain. However, once stained they are very resistant to destaining by vigorous methods. Several versions of the stain are used; the one given is the Ziehl–Neelsen method. A smear should be made in the normal manner and the following procedure used.

(a) The slide is covered with Ziehl–Neelsen's carbol fuchsin (see below) and heated (as in the spore stain) until steaming. It should not be allowed to dry out or boil. The heating should continue for five minutes.

(b) Wash with cold water.
(c) Decolourise with an acid/ethanol preparation of 3% HCl in 95% ethanol for 20 seconds.
(d) Counterstain with the methylene blue preparation used in the simple stain (above) for 30 seconds.
(e) Wash in water, blot dry, and examine.

Acid-fast organisms will give a red colour whilst non acid-fast organisms will give a blue colour.

Ziehl–Neelsen's carbol fuchsin may be prepared by dissolving 10 g of basic fuchsin in 100 ml of absolute ethanol and then adding this preparation to one litre of 5% phenol in water.

### 2.5 Lactophenol Stain for Fungi

Fungi may be stained using a lactophenol preparation. A sample of the fungus should be placed on a slide with a few drops of 95% ethanol, and teased out. The alcohol is allowed to dry and a few drops of a preparation of lactophenol are added. This is prepared by mixing 20 ml of distilled water, 20 ml of lactic acid, and 40 ml of glycerol and adding 20 g of phenol. When the phenol has dissolved, 0.075 g of methylene blue is added. The prepared slide is then covered with a coverslip which is pressed flat gently, and excess stain is removed from around the edge with blotting paper.

### 2.6 Other Stains

Numerous other stains are occasionally used in specialist microbiology laboratories. These include capsule stains, flagellar stains, negative stains, and specialist stains for protozoa and fungi. Details of these may be found in a specialist text book.[2]

## 3 REFERENCES

1. D.M. Green and S.S. Scott, 'Microscopy', in 'Mackie & McCartney Practical Medical Microbiology', Ch. 2, pp 11–37. 13th Edn. ed. J.G. Collee, J.P. Duguid, A.G. Fraser and B.P. Marmion, Churchill Livingstone, Edinburgh, London, Melbourne, and New York, 1989.
2. J.P. Duguid, 'Staining Methods', reference 1, Ch. 3, pp 38–63.

CHAPTER 5

# Microbial Detection and Counting

## 1 SAMPLING

The detection and identification of specific spoilage and pathogenic organisms is of paramount importance, but it is obviously not possible to test every item in a batch for quality, and batch sampling is therefore an integral feature of the test procedure. Equally, not all organisms likely to be present need to be individually detected, identified, and counted.

A typical approach would be to check a product for gross contamination and/or the presence of specific pathogens. This would give a useful guide to the condition of the produce and may suggest further tests.

There is currently no generally accepted method of examination, or any generally agreed safety standards and microbiological limits for most items. Some publications, such as the British Pharmacopoeia[1], list various tests, and any material sold with the BP label must have been subjected to these tests and conform to the limits given by the BP.

The sampling procedure used will obviously depend on the type of sample; whether it is liquid or solid; fresh, chilled or frozen; and the type of container (*e.g.* tinned, bottled). Other major problems are the frequency of sampling and the position on a production line from which a sample is taken. For example, when sampling from a food production line is carried out, an important consideration is whether or not the food has been subjected to sterilisation, or any form of pasteurisation after the point from which the sample was taken. This is considered further in the chapter on food microbiology under the concept of Hazard Analysis Critical Control Point (HACCP). Whatever the form of the sample, it should be collected in a sterile container using aseptic techniques, returned to the laboratory under conditions identical to those from which it was taken, and processed as rapidly as possible.

Homogeneous liquid samples present no serious problems, but if the liquid is non-homogeneous, *e.g.* milk, then the sample should be homogenised before sampling, as errors may arise if a non-standardised sampling protocol is used. Non-homogeneous solid samples may also contain micro-organisms confined to certain parts of the sample and again serious sampling errors may arise. If necessary, the material should be homogenised using a sterile blender or stomacher before the sample is taken. A series of sterile dilutions using Ringers solution or some other suitable diluent is then used to prepare a range of dilutions, which experience suggests will give counts of between 30–300 organisms per plate, when using the pour plate method (see page 45). These samples may then be used for both counting and identification.

One important point which should be made is that if a product is the source of food poisoning, or is microbiologically spoilt, then the microbial contamination is massive.

## 2 MICRO-ORGANISMS

The presence of certain bacterial species in items such as food, above certain numerical limits, suggests that the items may have been exposed to conditions which allow the growth of micro-organisms, and the possible proliferation of pathogens. In extreme cases the produce may be spoilt. The organisms present may be divided into four groups, essential, indicator, spoilage and pathogens, although there may be considerable overlap between the groups.

### 2.1 Essential Organisms

These are the organisms essential for making a product such as wine or cheese. These organisms are not normally counted, but it may be necessary to check on their numbers if there are problems with the production process.

### 2.2 Indicator Organisms

The most important tests for indicators, which give a microbiological 'feel' for the produce, are the Total Aerobic Viable (plate) Counts (TVAC), and a test for the presence of coliforms as indicators of faecal pollution. Other tests for indicators may be necessary for certain products or under certain storage conditions, *e.g.* psychrophilic plate counts for items stored at low temperature, anaerobic plate counts for suspect tinned items, and tests for enterococci in cases of suspected food poisoning. In certain items of produce, large numbers of indicator organisms may also pose a threat to health in their own right.

When indicator organisms are present, especially in high numbers, further tests for pathogens are imperative. However, the absence of indicators does not necessarily equate with microbiological quality and stability. Tests for the absence of specific pathogens and certain problematical spoilage organisms should therefore be included as part of the routine laboratory protocol. These may be run concurrently with tests for indicators and may be regarded as an early warning system of problems.

### 2.3 Spoilage Organisms

Tests and counting methods for specific spoilage organisms will vary widely depending on the type of produce (see Chapter 8). However, if a spoilage situation has already occurred or is imminent then the microbial count at the temperature of storage will be very high, necessitating a high dilution of the sample if significant counts are to be obtained.

### 2.4 Pathogens

High aerobic plate counts (TVAC) suggest that produce may be contaminated and further examination is essential. For example, positive coliform counts in food or water suggest more dangerous pathogens may be present, and that samples should be subjected to further rigorous examination. In the case mentioned, this would include tests for enteropathogenic coliforms, *Salmonella*, *Shigella*, *Staphylococcus*, and a variety of other pathogenic genera and species. The examination would involve suspect colonies from the preliminary tests (TVAC) and also fresh samples.

## 3 COUNTING

There are a wide variety of counting methods available and the one chosen will depend to some extent on the produce, the facilities available and the organisms one expects to find. One very serious problem is the time available, and obviously the more rapid the counting method the better. Many cultural methods take 1–15 days, which is totally unacceptable in a QC laboratory examining some bio-degradable product with a shelf life of only a few days. The time taken also depends on the number of organisms in the sample: a sample with $10^6$ organisms per gram will produce a response much faster than one with only ten organisms per gram.

Methods used for counting may be divided into three groups, (a) cultural or viable counts, (b) direct observation, and (c) measurement of some physical or chemical parameter. As a generality, the methods for counting bacterial numbers are considerably more advanced than methods for counting other microbial groups. The rapid counting and detection of fungi raises a number of serious problems. In addition, very little work has been carried out on the counting of fungi in the presence of bacteria. The detection and counting of pathogenic protozoa is extremely difficult. The counting of viruses such as enteropathogenic viruses is also very difficult and should be left to a specialist pathology laboratory. Although several of the methods for counting bacteria are very sensitive, no one method can be considered to be completely adequate for all requirements.

## 4 CULTURAL METHODS

These methods are by definition slow, taking from 24 hours to 10–15 days depending on the samples being examined. There is therefore considerable pressure to replace them with faster methods in the modern QC laboratory. They do, however, provide a very powerful amplification of microbial numbers, and are able to detect very low levels of contamination. Therefore these methods are still widely used and indeed are statutory requirements for the analysis of a number of products.

Another major advantage in addition to amplification is that cultural methods differentiate between viable and non-viable cells (*i.e.* those that can grow and those which cannot). The third advantage is that by use of selective media it is possible to differentiate between the various types of bacteria present. This may be essential in a situation where a pathogen is being overgrown by a non-pathogenic species. Cultural methods also have the major advantage that they are relatively cheap and are frequently the only realistic method available to developing countries.

There are disadvantages to cultural methods in addition to the long times required. From a practical viewpoint, a major problem is the large quantities of sterile pipettes, glassware and diluents required to dilute samples to give the requisite number of organisms on a plate. Further problems from a theoretical viewpoint include the problem of damaged cells which, although unable to grow, are still able to contribute to the modification of the produce. These cells are not counted by a cultural technique, even though they may contribute to spoilage problems.

The clumping of cells also causes difficulties. This occurs when cells aggregate into a clump which only forms one colony, and therefore leads to a considerable underestimation of microbial numbers. This problem has lead to the concept of colony forming units (cfu) rather than individual cells. This is a more serious problem with fungi than bacteria. Fungal colonies may arise from various types of spores or mycelium, and it is impossible to distinguish the origin of each colony. Each colony is therefore considered to be formed from a viable propagule (cfu), and the relationship between fungal biomass and the viable count is not always a direct one. Many plate counting techniques favour fungal species which sporulate strongly, and any disturbance of the plates causing fragmentation of the mycelium may give erroneously high results. This sporulation effect may be inhibited to some extent by the addition of Rose Bengal (Figure 5.1) to the medium. Antibiotics such as chloramphenicol (Figure 5.2) are also frequently incorporated into media when counting fungi to suppress the growth of bacteria, which normally grow more rapidly than fungi.

A number of cultural methods are available for counting. Most of them rely on prior dilution of samples.

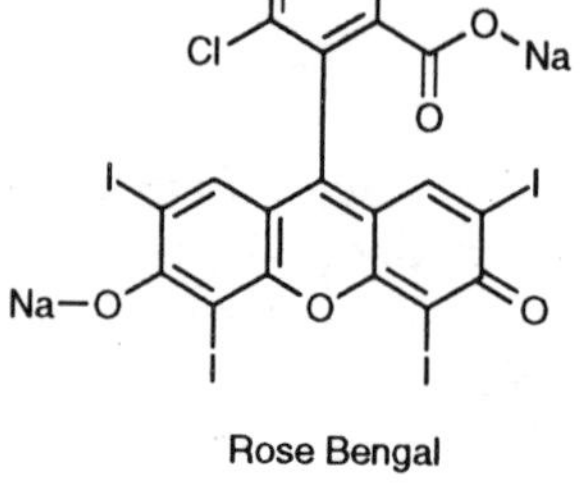

Rose Bengal

**Figure 5.1** *Structure of Rose Bengal*

## 4.1 Dilution of Samples

If the sample is a liquid, then 1 ml should be taken and added to 9 ml of a sterile diluent such as 1/4 strength Ringer's solution. This will give a dilution of one in ten ($10^{-1}$). This should be mixed thoroughly, 1 ml should be removed aseptically and added to a further 9 ml of sterile diluent to give a dilution of $10^{-2}$. This process should be repeated until a dilution of $10^{-6}$ is obtained. A suitable dilution should be chosen to give a count of between 30–300 organisms per plate when a plate count is carried out. The dilution on either side of this dilution should also be counted in case the estimate is either high or low. Thus, if it is estimated that a dilution of $10^{-4}$ is needed, counts should also be carried out on dilutions of $10^{-3}$ and $10^{-5}$ to give an adequate spread.

If the sample is a solid one, then 1 g should be suspended in 9 ml. of sterile Ringer's solution and homogenised thoroughly. This will give a dilution of $10^{-1}$, and the above procedure should then be followed to produce more dilute samples.

## 4.2 Dip Slides

These are microscope slides covered with a thin layer of nutrient agar or selective medium. They are dipped into liquid samples, the excess liquid is shaken off, the slide is transported to the laboratory for incubation and then may be counted under a microscope. They give overnight results and may be used in the field when immediate access to a laboratory is not possible. Due to the variations in the quantity of sample retained on the slide, they only give a rough indication of numbers and should always be supported by alternative counting methods.

## 4.3 Pour Plates

In this method 1 ml of sample or diluted sample is pipetted aseptically into a sterile petri dish, molten medium containing agar (usually 10 ml) is added, and the plate is rotated gently to mix the sample. The agar is allowed to set and the samples are incubated for the appropriate time. It is important that the agar is not too hot, or cells may be subjected to heat damage; agar at 55 °C is recommended. If the agar has been kept in a water bath, then the outside of the bottle should be wiped prior to pouring to remove excess water, which might otherwise contaminate the sample.

Roll tubes may be used as an alternative to plates. In these the sample is added to a tube containing molten agar which is then rotated rapidly on a mechanical roller until the agar sets. The tubes are incubated and counted. This method uses less agar than traditional pour plates and is widely used in the dairy industry.

## 4.4 Spread Plates

In this method, molten agar is poured into the petri dish, allowed to set and then dried thoroughly. The liquid sample is pipetted onto the plate and immediately

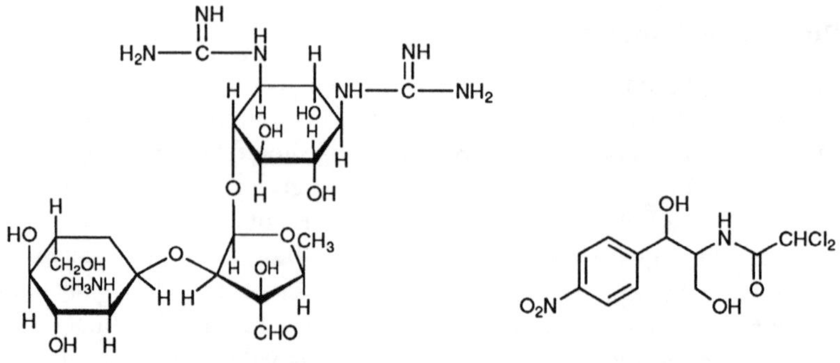

Streptomycin

Chloramphenicol

**Figure 5.2** *Two antibiotics which inhibit protein synthesis in procaryotic cells and which are widely used for suppressing bacterial growth in fungal media*

spread over the surface as quickly and uniformly as possible, using a sterile glass or plastic spreader shaped like a letter L.

Spread plates have certain advantages over pour plates for aerobic counts. The agar is not hot and therefore cells are not killed or heat damaged. Secondly, because all colonies are on the surface of the plate the availability of air to each colony is maximised.

Both spread and pour plates should be counted using a plate counter equipped with magnification, illumination, and a centimetre grid. A large number of electronic counters are available. The colonies on the plate should be counted promptly after the appropriate incubation period, or if this is not possible they should be stored in a cold room. If necessary, the edges of the plates may be sealed with cellotape to stop them drying out.

Plates containing 30–300 colonies should be counted and then multiplied by the appropriate dilution factor to give the microbial count of the original sample. A figure below 30 is not statistically accurate, whilst above 300 the colonies are so crowded that counting is impossible. It has recently been suggested that a count of 25–250 colonies per plate is better. If counts on diluted samples do not fall into the correct range, a fresh range of dilutions should be prepared. Generally dilutions in the range of $10^{-1}$–$10^{-6}$ will be adequate. With experience at handling certain types of samples, the range can usually be reduced considerably.

Media used for the counting of fungi using the spread or pour plate methods usually contain antibiotics to suppress the growth of bacteria (Figure 5.2). In the case of fungi which sporulate, Rose Bengal is also frequently incorporated to suppress sporulation which can lead to difficulties in counting.

### 4.5 Drop Plate (Miles and Misra) Technique

This method consists of allowing drops of a known volume from a calibrated pipette to fall from a predetermined height (2–3 cm) onto a plate of thoroughly dried agar. The plates are then incubated and counted, the total count being computed from the dilution factor and the volume of the drop. This method allows counts on the same sample to be repeated several times on a single plate, at a considerable saving in cost. The method is not suitable for species which form spreading colonies, and if these are anticipated then an alternative method should be used.

### 4.6 Automated Plate Counts

The spread plate technique may be automated by the use of a piece of apparatus known as the spiral plater. An agar plate is rotated on an Archimedean spiral whilst being inoculated. The volume of the sample decreases (and is therefore effectively diluted) as the spiral moves towards the outer edge of the plate. A specialised counting grid relating the area of plate to the sample volume enables colonies in the appropriate sector to be counted. An electronic colony counter travelling in the same Archimedean screw may be used.

This technique removes the need for multiple sterile dilutions, and can save a considerable amount of sample preparation and time when large numbers of samples are being processed. The apparatus is expensive and can only be justified where a large number of samples need processing.

### 4.7 Membrane Filtration

These methods may be used for liquid samples or solid samples which are soluble. They are especially useful for handling liquid samples containing only a small number of micro-organisms, when it becomes necessary to process large volumes.

The technique involves passing a large volume of liquid sample through a bacteriological membrane filter with a pore size usually of 0.45 μm. Once the sample has passed through the membrane filter, the membrane is removed and may either be stained and subjected to direct counting, or placed on a growth medium. A fluorescent dye may be used if direct counting is chosen. Alternatively, the membrane may be placed on a layer of suitable medium, incubated for the appropriate time and counted. Once the cells have been counted, this can be related to the volume of the original sample. The medium used may be selective, thus enabling a specific species or genus to be assessed. The method can be used for both bacteria and yeasts.

A number of companies sell the equipment and also sterile dehydrated pads of selective media for use with membrane filters. These only need the addition of sterile water before use. The most important suppliers in the UK include companies such as Gelman, Millipore, and Sartorius.

One problem common to all cultural techniques, whether plate counting or membrane filtration, is the degree of recovery of damaged organisms.

## 5 DIRECT COUNTING METHODS

In some cases it is possible to see microbial contamination. Materials such as cereal grains and oil seeds can be examined for surface fungi directly, either before or after a period of incubation. In severe cases of *Fusarium* infections, cereal grains may be a red/pink colour.

The direct counting of bacteria and moulds has been widely used. It consists of spreading a known volume or weight of produce onto a slide containing a grid covering a prescribed area. The slide is then fixed, defatted if necessary, and stained. It is also possible, using a slide with a millimetre scale on it, to work out the area of a microscope field under both low and high powers. The number of cells in a given number of microscope fields can then be counted and related to the number of organisms per gm or ml. This method may also be carried out using a slide with a counting chamber (Helber chamber) holding a known volume of liquid.

Direct counting has a number of advantages, it is rapid, cheap, and may be done concurrently with a Gram stain. There are also serious disadvantages, which include the problem of distinguishing debris from micro-organisms, the inability to distinguish viable and non-viable cells, operator time and fatigue, and poor reproducibility.

Direct counting can be improved by the use of fluorescent dyes, such as acridine, especially if combined with the recovery of cells by membrane filtration. Direct epifluorescent filter techniques (DEFT) are used in the milk and dairy industries to estimate both bacteria and fungi.[2] They can produce results in less than 25 minutes which correlate closely with traditional methods. Further developments have automated the counting procedure by the use of image analysers thus removing the problem of operator fatigue.

In addition to fluorescent stains, fluorescent antibody techniques may be used. Such methods have been used to study micro-organisms in a number of commodities, and in many cases the results are claimed to correlate closely with cultural methods. The methods involve using a fluorescent chemical tagged with a specific antibody. The antibody recognises a specific micro-organism and attaches to it, thus tagging the micro-organism with a fluorescent dye. This method not only detects the organism but also identifies it. The major advantage of fluorescent methods is the very low detection limits, but non-availability of specific fluorescent labelled antisera limits the immunological aspects of these techniques severely.

One type of direct counting method which has been standardised is the Howard mould count[3] which is widely used by the Food and Drug Administration in the USA.[4] Although widely used, the method is inaccurate, subjective, and shows poor reproducibility.

## 6 PHYSICAL AND CHEMICAL METHODS

A number of methods have been used, but many are confined to a small number of research laboratories and only those methods in general use will be discussed.

### 6.1 Spectrophotometric Methods

The absorbance or light scattering properties of a microbial culture are related to the number of organisms present. Calibration curves can be prepared of absorbance against numbers using some viable counting method. These can then be used to estimate the numbers of organisms present in broth cultures. If the appropriate tubes have been used for growing the organisms, samples can be read directly in a nephelometer using commercially available opacity tubes as standards. Calibration curves need to be prepared for each species of bacteria as the light scattering properties vary from species to species.

## 6.2 Dry Weight

There is a close relationship between dry weight and microbial numbers. Although this method would give a good measurement of micro-organisms, there are serious problems in separating the micro-organisms from the sample, and the preparation time of samples is excessive.

## 6.3 Impedance and Conductivity

Physical methods include the use of electrical impedance and conductivity.[5] These are based on the resistance to the flow of an electric current through the medium. This changes as bacteria grow and convert non-electrolytes such as sugars, into electrolytes such as lactic and acetic acids. The impedance remains relatively constant until the number of cells reaches a threshold ($10^6$–$10^7$ $ml^{-1}$), when marked impedance changes occur. The time taken to reach the threshold value is the impedance detection time, and can be related to the number of organisms in the original sample. Although the apparatus is good at measuring high numbers of organisms rapidly, it becomes less reliable at low levels of contamination. Selective media may be used in impedance systems which allows the estimation of specific species, although the number of selective media available for fungi is not high, which restricts its use for these organisms.

The method has been used for detecting both bacteria and fungi, and is finding increasing use in industry due to its speed, but the cost of the apparatus is high, and can only be justified if there are a large number of samples. The use of this method in the developing world is unlikely due to the cost.

## 6.4 Particle Counters

Electronic particle (Coulter) counters have been used for counting bacteria and yeasts. Major users are in the wine and beer industries. The problems with this method are that it is non-specific, and there are considerable difficulties in preparing samples which are free of dust and detritus as these may cause erroneous high counts.

## 6.5 ATP

The use of adenosine triphosphate (ATP) (Figure 5.3) to detect live cells depends on the use of a firefly luciferin reaction detected by scintillation counting.[6] This

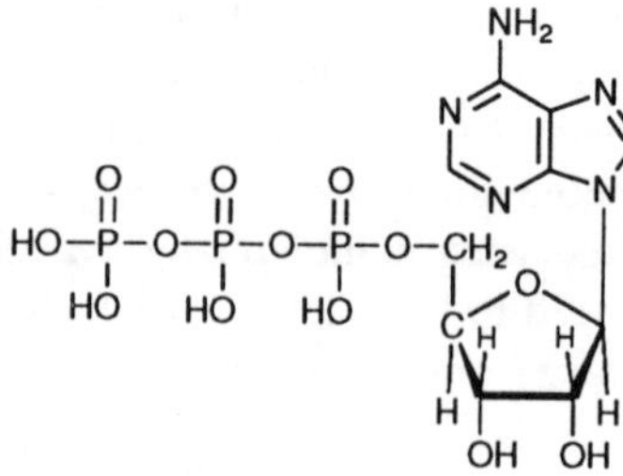

**Figure 5.3** *Structure of adenosine triphosphate (ATP)*

method has been used for rapid estimation of the microbial contamination of raw meat, but does not distinguish between fungi and bacteria. It depends upon being able to separate the ATP derived from microbial contamination and the ATP from the background material. The limits of detection are reported to be of the order of $10^3$ organisms per gram.

Although rapid and sensitive, ATP detection is very expensive. The cost and level of operator competence required are such that the method cannot be justified unless a large number of samples are envisaged.

A variation on this theme consists of using a genetically engineered bacterial virus (phage) containing the luciferase gene. If this is added to a sample then the virus would attack certain specific bacteria and introduce the luciferase gene into them. The bacteria containing the gene would scintillate in the presence of ATP and could be estimated. The method would be specific for those bacteria attacked by a specific virus, and could be altered by changing the viral carrier.

## 6.6 Mycotoxins

The detection of mycotoxins is of considerable importance in food production, but there are a number of criticisms in using this as a method of detecting fungi and fungal spoilage, although it is widely used. Mycotoxins are secondary metabolites only produced after the fungus has been growing for some time, and toxins are therefore indicative of damage which has already taken place.

There is no correlation between toxin production and cell numbers for a variety of reasons. The conditions under which different fungi grow and form toxins are not identical, and there are frequently a succession of toxins, some of which may be modified or metabolised by the organisms producing them or by other species. Toxin production has also been reported to depend on a number of parameters, which include temperature, water, time of incubation, oxygen, carbon dioxide, nitrogen source, and presence of trace metals. Finally there are a number of non-toxigenic fungi which would not be detected by these methods. It has been shown that there are a significant percentage of *Aspergillus flavus* strains which do not produce aflatoxins. Such fungi, however, are still capable of producing physical damage to produce of the type specified in Chapter 2.

However, although these comments are obviously of relevance, the detection of toxins is of great importance as many toxins are known carcinogens. Over two hundred mycotoxins have now been described which fall into a number of groups. Details of several of the more important groups are given below.

The aflatoxins are probably the best known group and also the most toxic (Figure 2.2). They are formed by the species *Aspergillus flavus* and related organisms. The most common and probably the most toxic is aflatoxin $B_1$ which is a very potent liver carcinogen, and has been linked epidemiologically to liver cancer in a number of developing countries. Hydroxylation of this molecule gives rise to aflatoxin $M_1$ which is usually found in milk from cattle fed with contaminated feed. Aflatoxins have been found in numerous foodstuffs of which mouldy peanuts are the most common example, and which have been linked to a number of outbreaks of aflatoxicosis in both humans and farm stock.

The ochratoxins are formed by members of both the Aspergilli and Penicillia and are nephrotoxins damaging the kidneys and causing renal disease. Patulin is produced by numerous members of the Aspergilli and Penicillia and is found widely in a number of commodities, especially apple juice. It is known to be both mutagenic and carcinogenic. The trichothecenes are produced by numerous fungi but the best known example is caused by *Fusarium* spp. growing on cereals. In extreme examples the ears of cereal may be coloured pink. The two best known toxins in this group are deoxynivalenol (vomitoxin), which produces serious stock losses, and diacetoxyscirpenol ($T_2$ toxin). This latter toxin is believed to have caused widespread fatal outbreaks of alimentary toxic aleukia due to the ingestion of mouldy grain in Russia in the last century and also in 1944 during the siege of Stalingrad. Examples of some of these toxins are shown in Figure 5.4.

The determination of mycotoxins has been reviewed by Jarvis *et al.*[7] and van Egmond.[8] Numerous methods have been described, including both biological and chemical assays with detection levels ranging from sensitive to relatively insensitive.

The minicolumn methods of Holaday and Romer[9] are ideal for developing

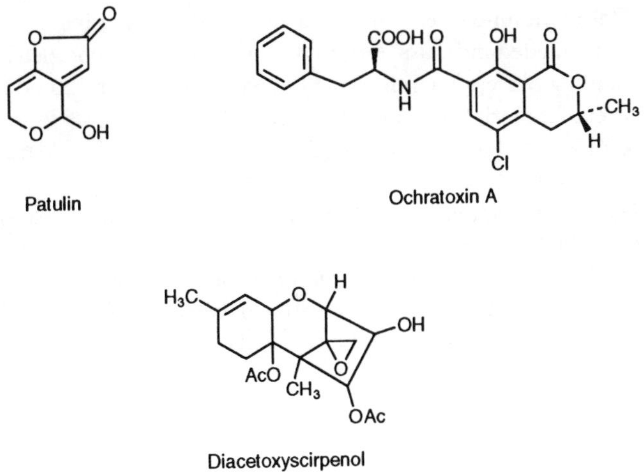

Patulin

Ochratoxin A

Diacetoxyscirpenol

**Figure 5.4** *Examples of mycotoxins*

countries, being rapid, cheap, and not requiring sophisticated equipment. However, they do have higher limits of detection than methods such as thin layer chromatography (TLC), high performance liquid chromatography (HPLC), enzyme-linked immunosorbent assays (ELISA) and radioimmunoassays (RIA).

Two dimensional TLC probably offers the best rapid, cheap method of detecting many toxins, especially in countries where the purchase and maintenance of expensive instruments may not be possible. It also has the major advantage in many developing countries that it is not reliant upon a source of power.

GLC of mycotoxins is limited as most are non-volatile and derivatives need to be prepared, although GLC is of considerable use in the determination of the trichothecenes which are difficult to assay by TLC or HPLC. The preferred method of detection is electron capture, but increasing use is being made of GLC linked to mass spectrometry.

Numerous HPLC methods have been developed for toxin determination and van Egmond has reviewed these.

Enzyme-linked Imunosorbent Assay (ELISA) and Radioimmunoassay (RIA) are highly specific methods that have been used for the assay of a number of mycotoxins. ELISA is the preferred method of immunoassay because of the limited half life of the radioisotopes and the problems of disposal of radioactive waste.

Mycotoxins are low molecular weight organic compounds which do not possess any antigenic properties. However, if they are conjugated with protein such as Bovine Serum Albumen (BSA), they become antigenic and can be used to prepare specific antisera.

Basically in ELISA and RIA, an enzyme is linked to an antigen in such a manner that the activity of the enzyme is altered when the antigen combines with with antibody. The assay system consists of enzyme labelled antigen, antibody, enzyme substrate and test sample. If the test sample contains any antigen, it combines with the antibody leaving the enzyme–antigen complex free to attack the substrate. Radioimmunoassay works in a similar manner using a radioactive indicator. If the unlabelled indicator is present in the sample, it competes with the radiolabel for binding places. Therefore the higher the level of radio bound label, the lower the level of material present in the sample.

Both methods have a sensitivity of the order of 1 $\mu g\ kg^{-1}$ for aflatoxins, but procedures for other mycotoxins are only being developed slowly.

Occasionally, an organism may produce some metabolite which reacts rapidly. A differential medium based on ferric citrate, which allows a presumptive identification of *Aspergillus flavus* and *A. parasiticus,* utilises the production of the toxin aspergillic acid by these species. Improved versions of this method give results in 42 hours which are comparable with plate counts on standard media taking 5–7 days.

## 6.7 Limulus Lysate (LAL)

Bacteria also produce toxins which have traditionally been tested by the use of

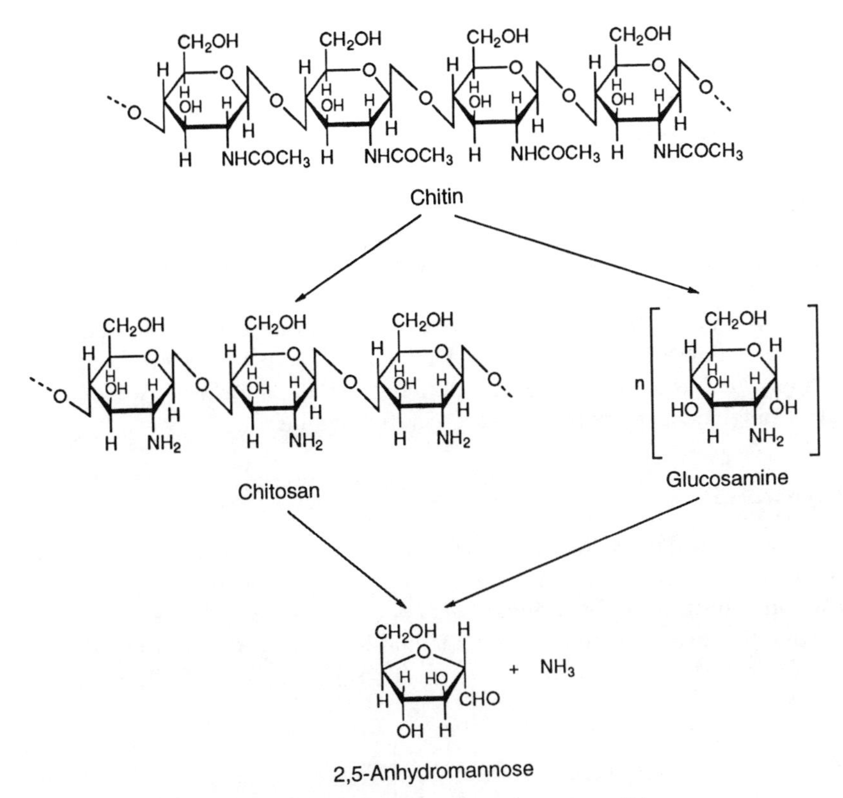


**Figure 5.5** *Scheme showing formation of 2,5-anhydromannose from chitin*

animals. The Limulus Amoebocyte Lysate (LAL) test detects bacterial endotoxins of the type formed by Gram-negative bacteria.[10]

The horseshoe crab (*Limulus*) suffers from a disease caused by the genus *Vibrio*, which causes fatal intravascular clotting. This observation led to the realisation that the amoebocytes collected from the haemolymph of the crab contained a reactive protein (known as LAL), which could be released by lysis in distilled water. The LAL reagent is specific for the endotoxin of Gram-negative bacteria. This endotoxin is the lipopolysaccharide (LPS) fraction, and it is the lipid portion which reacts with LAL.

Several assay methods have been used. These include gelation, of which there are several variations, turbidity and a test involving the measurement of released chromogenic material, which is the most widely used. This is more rapid and sensitive than conventional gelation tests and is more easily automated. A chromogen such as *p*-nitroaniline is attached to a synthetic peptide, the chromogen is cleaved off during the course of the reaction and can be measured spectrophotometrically. The chromogenic test is capable of measuring endotoxin at levels of $> 5 \text{pg ml}^{-1}$.

Problems may arise as the LAL from different manufacturers is variable and also the LPS from different Gram-negative bacteria does not produce an equal response. It is obvious that there cannot be a direct conversion of LPS into cells, and standardisation of methodology is therefore of considerable importance. However the method is extremely sensitive and claims of detecting as few as 100 *Pseudomonas aeruginosa* and 500 *Escherichia coli* have been made.

## 6.8 Chitin

Fungi produce the polysaccharide chitin as part of their cell wall. This is a polymer of *N*-acetyl glucosamine which can be deacetylated to form chitosan. The chitosan can be treated with nitrous acid to form 2,5-anhydromannose which can be estimated colorimetrically (Figure 5.5). The test takes five hours but is not widely used as any insect debris interferes with it, and it is not suitable for estimating fungi in any plant products high in sugar amines.

## 6.9 Ergosterol

Ergosterol (Figure 5.6) is found in the cell membrane of fungi and can be estimated either by HPLC or by GLC after conversion to a silane derivative. The test takes one hour, is specific to fungi but is not linear under all conditions. It is possible that there is some future for this method but further work is needed.

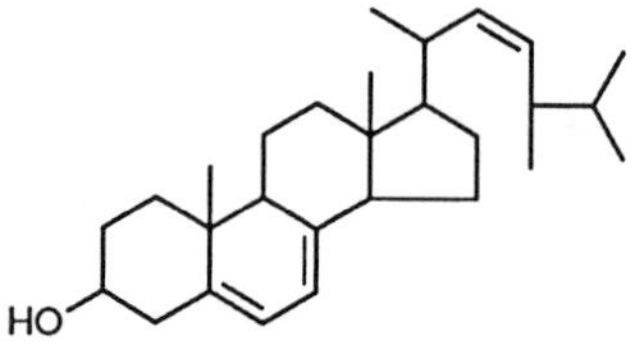

HO

**Figure 5.6** *Structure of ergosterol*

## 6.10 DNA Probes

A method being developed increasingly is the use of DNA probes. These use radiolabelled single stranded DNA (the probe), which attaches to complementary DNA to form double stranded DNA hybrids. The test is carried out on a solid phase such as a cellulose nitrate filter. The double stranded DNA remains bound to the filter, whilst the single stranded material is washed off. The radioactivity taken up into the double strand may then be assessed. The method is rapid and has great potential, although the main drawback will probably be the disposal of the radioactive material.

## 6.11 Monoclonal Antibodies

Monoclonal antibodies are also becoming increasingly available, and have great potential in the rapid detection of micro-organisms. The use of this method at present is limited by the small number of antibodies available.

There are other chemical methods used for the estimation of fungi, but they are not in widespread use and are rarely found outside research laboratories.

## 7 CONCLUSIONS

The counting of micro-organisms raises a number of serious problems. These include:

(a) Sampling, especially of heterogeneous systems.
(b) Estimation of viable as opposed to non-viable organisms if direct observation is used.
(c) The time factor and choice of media if a cultural method is used.
(d) The separation of micro-organisms and background material if a chemical method is used.
(e) The measurement of any toxic material is only indicative of damage which has already taken place.
(f) The frequently non-linear relationship between micro-organism and any parameter measured.

As a general statement, microbiologists are fairly jaundiced about rapid methods replacing traditional counting methods, and industry broadly speaking has never produced a technology which answers all criticisms. The cost and maintenance of equipment is also of importance, and these factors are most likely to be at a premium in those parts of the world where damage from spoilage micro-organisms is greatest. All these factors must be taken into account when a decision is made to use a particular method, and it must be accepted that any decision can at best only be a compromise solution.

## 8 REFERENCES AND FURTHER READING

1. British Pharmacopoeia, British Pharmacopoeia Commission, HMSO, London, 1988.
2. G.L. Pettipher, R.G. Kroll, L.J. Farr, and R.P. Betts, in 'DEFT: Recent Developments for Food and Beverages', Rapid Microbiological Methods for Foods, Beverages and Pharmaceuticals', Society for Applied Bacteriology, Vol. 25. ed. C.J. Stannard, S.B. Petitt and F.A. Skinner, Blackwell Scientific Publications, 1989.
3. B.J. Howard, US Dept of Agriculture, Bureau of Chemistry, Circular No. 68, 1911.
4. Bacteriological Analytical Manual, 6th Edn. Food and Drug Administration, 1984, Association of Official Analytical Chemists.
5. M.C Easter and D.M. Gibson, Detection of Micro-organisms by Electrical Measurement, *Progress in Industrial Microbiology*, 1989, **26,** 57.
6. C.J. Stannard , ATP Estimation, *Progress in Industrial Microbiology*, 1989, **26**, 1.
7. B. Jarvis, W.B. Chapman, D.M. Norton, and G.M. Toule, 'Methods for the Detection and Identification of Selected Mycotoxins', in 'Isolation and Identifica-

tion Methods for Food Poisoning Organisms', ed. J.E.L. Corry and F.A. Skinner, Society for Applied Bacteriology Series Vol. 17, Academic Press, London, 1982.
8. H.P van Egmond, 'Determination of Mycotoxins', in 'Developments in Food Analysis Techniques,' ed. R.D. King, Vol. 3, 1984.
9. C. E. Holaday, 'Rapid Method for Detecting Aflatoxins in Peanuts', *J. American Oil Chemists Society*, 1968, **45**, 680.
10. J.M. Jay, 'The Limulus Amoebocyte Lysate (LAL) Method', *Progress in Industrial Microbiology,* **26**, 101.

CHAPTER 6

# Biochemical Testing

The biochemical processes referred to as metabolism can be divided into primary and secondary metabolism. All organisms possess similar metabolic pathways by which they synthesise and utilise certain essential chemicals, *e.g.* sugars, amino acids, nucleotides, fatty acids, and the polymers derived from them. These molecules, which are essential for the metabolism of the cell, are called primary metabolites.

In addition to the primary metabolites, there are also compounds produced by other metabolic pathways which have no apparent necessity or use in the cell. These are secondary metabolites produced by pathways of secondary metabolism and are frequently only formed when the cell ceases active growth.

The dividing line between primary and secondary metabolism is somewhat blurred in many cases. For example, amino acids are generally considered to be primary metabolites, but some are sufficiently obscure to be considered as secondary metabolites, whilst sterols which are frequently considered as secondary metabolites are essential to the cell membrane of eucaryotic species, and therefore must be regarded as primary metabolites. Generally, the primary metabolic processes are too similar from group to group to be of much use in identification, but the secondary metabolites which tend to be unique to a small group are used widely.

The initial problem before commencing identification is to obtain a bacterial isolate in pure culture, and before beginning any biochemical tests the purity of the cultures should be confirmed by plating out onto solid non-selective media. Selective media should not be used as many selective media only suppress the growth of unwanted organisms; they do not kill them. In the case of highly contaminated sources, such as faeces, the isolation and purification procedures may take several days. Methods for doing this are discussed in Chapter 3.

The use of growth independent enzyme assays (GIEAs) is increasing and forms the basis for many of the rapid identification kits now available. Rapid methods of identifying a number of genera and groups are available. These include enterobacteria, anaerobes, staphylococci, streptococci, and yeasts. These methods may give results in four hours using computerised data bases. One point to be stressed, however, is that although these tests are rapid, it is still essential to obtain a pure culture of the organism to be tested, and therefore the improvements claimed by manufacturers in the time taken to identify an organism are frequently not as rapid as implied in their literature.

Two varieties of micro methods are used in rapid testing. In the first method, dehydrated substrates in plastic compartments are provided in a strip. These would include all the tests necessary to differentiate an organism known to be, for example, a yeast. The strip of tests is inoculated with a pure culture of organisms and incubated for several hours. The results are then scored as positive or negative, and either checked manually against tables or run against a computerised data base. The result is usually expressed as a probability, *i.e.* the probability of this organism being X is better than 95%. There are several of these systems available but the best known is probably API, which is widely used and has an excellent reputation for reliability and customer service.

The second type of system is based on agar. An example of this is the Roche Enterotube. The main problem of these systems is that the agar dries out and the shelf life is limited. The tubes consist of a number of compartments, each containing the ingredients for a different test. An inoculating wire runs through the tube and all the tests. The tube is inoculated by touching one end of the wire to a colony and drawing the wire through the tube, thus inoculating each test in turn. The tubes are incubated overnight and the results compared with tables. The tubes are expensive, but useful in a laboratory with limited facilities.

The reproducibility of the test is of fundamental importance if it is to be used reliably. Standardised methods should be used and controls using known positive and negative samples should be included. Generally the organisms causing the greatest problems are those giving a weak positive reaction, and even the experienced microbiologist may have problems making decisions. The use of GIEAs removes the errors associated with the type of test that depends on the growth rate, as fluctuations can be caused by the time of reading, depending on whether the organism is slow or fast growing.

One of the major problems is that not all members of a genus may give the same result, and tables frequently quote measurements such as 85% positive for a certain test. This can cause difficulties in recognising unusual strains.

Frequently the only quantitative measurements used by microbiologists are time and temperature. Generally, very little attention is paid to aspects such as the size of inoculum, volume of test medium, or the age of the culture from which the inoculum is made.

A number of points cause problems. Growth is frequently assessed visually by examining for turbidity. This may cause problems due to the precipitation of insoluble material from the medium. A common problem is magnesium phosphate. Growth media contain phosphate, usually added as one of the potassium salts, and also magnesium added as the sulfate. Autoclaving this combination may result in a precipitate of magnesium phosphate. This can be avoided by autoclaving the magnesium salt separately from the rest of the medium, and mixing when cool.

Problems may also be caused by excessive autoclaving of media high in sugars. This may cause caramelisation and can be avoided by sterilising sugar solutions by membrane filtration and adding to the rest of the medium when cold, to give the required concentration. This avoids the build-up of furfural derivatives which are toxic to many micro-organisms.

Carry over of excessive media from the inoculum into the test system may be another source of error. This can be avoided by inoculating material from a presumptive positive test into a fresh tube of test media and reading a second time.

The incorporation of pH indicators into the medium may also create problems. These fall into several areas.

(a) The redox potential of the medium may change in the relatively anaerobic growth conditions of many cultures. This can cause irreversible bleaching of some indicators.
(b) Some indicators are antibacterial.
(c) Indicators cannot be used in highly coloured media or in cultures where the cells develop pigments.

Some tests involve adding an indicator to the culture after it has grown. Most microbiologists usually add a few drops of indicator to the culture, but it is better practice to place a few drops of culture medium onto a tile with the indicator.

A wide variety of biochemical tests can be applied to micro-organisms for the purpose of identification and these fall into several groups. It is unusual for any one test to give a definitive answer, and usually several tests are required, some of which may be carried out simultaneously.

Although biochemical tests can be broken down into various groups, the tests are a continuous spectrum and many areas overlap, *e.g.* the presence or absence of a specific enzyme may determine whether or not a compound can be used as a carbon source. The groups are frequently presented as shown below:

(1) Carbon and energy source
(2) Oxygen requirements
(3) Temperature
(4) Metabolic end products
(5) Enzymes
(6) Nitrogen requirements
(7) Decomposition of large molecules
(8) Miscellaneous

The remainder of this chapter contains examples of each type, and although it is not comprehensive, it contains the tests used to separate the more common species.

## 1 CARBON SOURCE

The ability of an organism to grow using citric acid as sole carbon source is widely used as a test for differentiating certain members of the group Enterobacteriaceae. Two media commonly used are Kosers citrate medium in which growth is shown by turbidity, and Simmons citrate agar which contains Bromothymol Blue (Figure 6.1) as indicator and growth is indicated by a blue

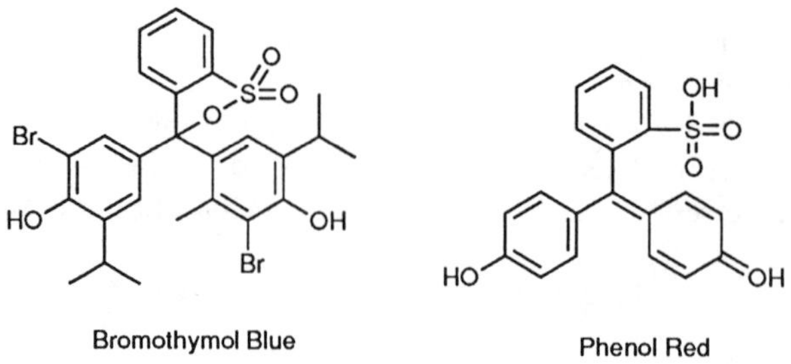

Bromothymol Blue

Phenol Red

**Figure 6.1** *Structures of Bromothymol Blue and Phenol Red*

colour. In the case of non-growth the medium remains green. Both Kosers citrate and Simmons citrate use inorganic nitrogen sources. A variation is Christensens citrate medium which uses an organic source of nitrogen and Phenol Red (Figure 6.1) as indicator with growth being indicated as a deep red colour. This Christensens medium should not be confused with the Christensens medium associated with the test for urease production. Growth on citrate is used mainly as a method of differentiating coliforms, *Escherichia coli* gives a negative result.

Growth on malonic acid as sole carbon source is used less commonly as a differential test for *Enterobacter* and *Escherichia*. Growth is shown by the indicator (Bromothymol Blue) changing to its alkaline colour, Prussian Blue.

When carrying out these tests it is essential to avoid carry over of medium in the inoculum, and it is best to carry out a second definitive test after obtaining a presumptive positive in the first test. A loopful of organisms from the first positive test (presumptive) is used as inoculum for the second test (confirmatory).

The use of sugars as sole carbon source is dealt with under oxygen requirements and end products.

## 2 OXYGEN REQUIREMENTS

This tests the ability of the organism to use compounds such as sugars or sugar alcohols either oxidatively or fermentatively. This is one of the most important tests in microbiology and is used to distinguish between various groups of Gram-negative organisms and also the Gram-positive Micrococcaceae.

The test can easily be carried out by a stab culture into a tube of test agar containing the required carbon source and the appropriate indicator. Anaerobic bacteria change the colour of the medium at the bottom of the tube by converting sugars to acids. Aerobic bacteria change the colour of the indicator at the top. Facultative organisms change the colour throughout the tube. However, if tubes are left too long before reading, it may become difficult to determine which is which, due to the diffusion of acid. Therefore, tubes are frequently set up in duplicate and a layer of sterile liquid paraffin is placed on top of one tube, thus accentuating the anaerobic reaction. Oxidative organisms will produce acid only in the tube open to the air, facultative organisms will produce acid in both tubes, whilst anaerobes only form acid in the tubes sealed with oil.

If the agar in the above test is semi-sloppy, then the test may, in competent

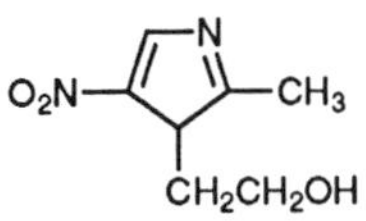

**Figure 6.2** *Structure of the drug metronidazole*

hands, also be used to detect gas formation (bubbles appear in the agar). Some authors claim that it can also be used to detect motility by noting whether growth has occurred some distance from the stab line.

Liquid media may also be used for testing oxidative/fermentative reactions (Hugh and Leifson's O/F test), and if this is done then a small inverted test tube known as a Durham tube is added. This allows the reaction to be examined for the formation of gas as well as acid. Tubes are again prepared in duplicate, and sterile liquid paraffin or mineral oil is again added to one tube to accentuate the anaerobic reaction. A few organisms may produce an alkaline reaction due to the conversion of peptones in the medium to amines.

One useful method of testing for anaerobic organisms is that all anaerobes are sensitive to the drug metronidazole (Figure 6.2), and are incapable of growth in the presence of this compound.

## 3 TEMPERATURE

Organisms may be divided into three groups depending on their temperature requirements, psychrophiles, mesophiles, and thermophiles. The division is arbitrary, but is generally regarded as below 10 °C, 10–45 °C, and above 45 °C.

There are a number of reactions used in identification for which the temperature is critical, *e.g.* the growth of thermotolerant coliforms at 44 °C in various media is used as a diagnostic test, especially in water samples. Other species in which temperature is important are those which can grow at low temperatures in refrigerators, thus causing problems such as spoilage and food poisoning, *e.g. Listeria.*

## 4 DETECTION OF SPECIFIC METABOLIC END PRODUCTS

### 4.1 Methyl Red Test

This determines whether the organism can produce sufficient acid from glucose to reduce the pH to 4.5 or below and maintain it there. It is used for the differentiation of Enterobacteria.

### 4.2 Voges–Proskauer Test

Many organisms impose specific end reactions on the Embden Meyerhof pathway and this test is designed to detect the presence of 2,3-butandiol, acetoin (3-hydroxy-2-butanone) or diacetyl (2,3-butanedione). Aeration of all these compounds in the presence of KOH produces diacetyl which reacts with peptones in the media to produce a pink colour. Tubes need vigorous shaking and the maximum colour may not develop for an hour or so. The test is used for differentiating Enterobacteria.

The Methyl Red and Voges–Proskauer tests are widely used to distinguish between *E. coli* and *Klebsiella* species with *E. coli* being MR positive, VP negative and *Klebsiella* being MR negative, VP positive.

### 4.3 Indole Production

This test consists of incubating cells in a medium containing the amino acid tryptophan and measuring the production of indole. A culture 2–3 days old is added to Kovacs reagent (dimethylaminobenzaldehyde in HCl) and a pink/red colour gives a positive reaction. *E. coli* gives a positive result, and the test is frequently used to distinguish it from *Klebsiella.*

The combination of Indole, Methyl Red, Voges–Proskauer and Citrate tests is frequently known under the acronym IMViC.

### 4.4 $H_2S$ Production

Many organisms are variable in their ability to produce hydrogen sulfide from sulfur containing compounds. It is essential when reporting $H_2S$ production to define the medium used. Many weakly positive organisms produce $H_2S$ from media rich in sulfur compounds when an indicator such as lead acetate, which is very sensitive, is used. Filter paper impregnated with lead acetate can also be used and is placed at the top of the slope or test tube. Blackening of the filter paper or medium indicates a positive reaction. However, any organisms producing $H_2S$ from a medium low in sulfur compounds, in the presence of ferrous chloride, which is a poor indicator, must be a very strong $H_2S$ producer. This allows distinction between various Enterobacteriaceae. The method must therefore be reported, especially when dealing with a genus such as *Salmonella.*

Two media commonly used to demonstrate $H_2S$ production are Kliglers Iron Agar and Triple Iron Agar. The rationale of these two media is as follows. Kliglers Iron Agar contains two sugars, lactose at high concentrations and glucose at low concentrations with phenol red as indicator of acid production, and ferrous sulfate as indicator of $H_2S$ production. Organisms capable of utilising glucose but not lactose will produce an acid reaction over a short time period. On the surface of the slope this will change to alkaline as the acid products of glucose metabolism are further oxidised. Where oxygen is excluded in the butt of the slope the reaction will remain acid. Thus, organisms which are glucose utilisers and lactose non-utilisers will produce an initial acid reaction throughout the slope which will change to produce an alkaline reaction at the surface with an acid butt. Lactose utilisers will produce an acid reaction throughout which will persist. A blackening of the medium will indicate an $H_2S$ producer, and production of gas will be shown by bubbles forming in the agar.

Triple Iron Agar works in a similar manner but contains three sugars, low concentrations of glucose, high concentrations of lactose and sucrose, Phenol Red and ferrous sulfate.

Both of the above media should only have their caps screwed on loosely after inoculation to allow the ingress of air, thus enabling the alkaline reaction to be produced. In both cases it is necessary to confirm the biochemical results in single sugar media.

## 4.5 Fatty Acids

The production of volatile fatty acids (VFA) from formic to heptonic acids is a stable characteristic both qualitatively and quantitatively, and allows the differentiation of certain species in the group Bacteroides. The formation of branched chain fatty acids may also be used for differentiation. The use of this property would only be found in a specialised laboratory.

## 4.6 Acetone/Isopropanol/Butanol/Butyric Acid

The presence of these may be used in the differentation of *Clostridium* spp. They are detected in the medium by means of GLC. Testing for these compounds is not used on a routine basis and would only be found in laboratories specialising in clostridia.

## 4.7 Growth on Sugars

Growth on a wide variety of sugars and sugar alcohols is widely used as a secondary test for differentiating many species. The compounds used include glucose, lactose, sucrose, ribose, arabinose, rhamnose, mannitol, dulcitol, salicin, adonitol, inositol and sorbitol. An indicator is added to the medium and utilisation of the sugar to form acids is shown by a change in colour.

The medium may be solid or liquid. In the case of liquid medium, a Durham tube may be added in order to detect any gas which may be formed. The difference between acid, and acid plus gas may be of diagnostic value.

A number of different media are available and failure to standardise the tests causes confusion. Peptones in the medium break down to form alkali and if the carbohydrate is metabolised to form acid, then the acid reaction is only seen if the acid is predominant. Therefore a medium high in peptone may mask a weak acid reaction. Different indicators are also used, bromothymol blue changes colour at pH 6.0 whereas bromocresol purple changes at pH 5.0. Peptone water is commonly used in the UK. Some aerobes such as *Pseudomonas* give unreliable results on a peptone medium and need to be tested on medium containing ammonium sulfate as a nitrogen source.

A commonly used medium of this type is MacConkey broth. This contains the sugar lactose, the bile salt sodium taurocholate and an indicator. The bile salt only allows the growth of bile tolerant species *i.e.* enterobacteria, and metabolism of the sugar is shown by the acid colour of the indicator. Lactose is one of the most important sugars in these types of media, and is used to distinguish between the pathogens *Salmonella* and *Shigella* which are lactose negative, and genera such as *Escherichia* and *Klebsiella* which ferment lactose rapidly. However, *Salmonella arizona* and certain *Shigella* species are able to ferment lactose slowly (late lactose fermenters) and may be distinguished from other species by their reaction to sugars other than lactose.

## 5 ENZYMES

There are numerous tests for a wide variety of enzymes and only the more commonly used ones will be discussed.

### 5.1 Urease

This enzyme splits urea to ammonia and carbon dioxide (Figure 6.3). The carbon dioxide is released and the ammonia raises the pH of the medium turning any indicator present alkaline. The test is used widely to distinguish *Proteus* species from other Gram-negative rods. A heavy inoculum is used and a positive result is frequently obtained within 2–3 hours.

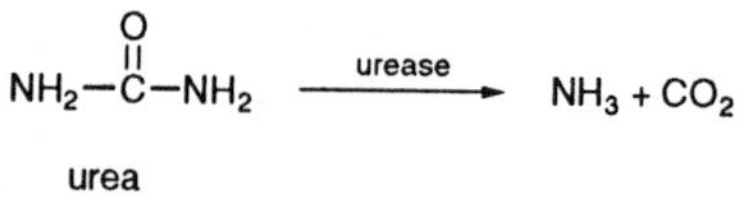

$$\underset{\text{urea}}{NH_2{-}\overset{\overset{\large O}{||}}{C}{-}NH_2} \xrightarrow{\text{urease}} NH_3 + CO_2$$

**Figure 6.3** *Schematic representation of urease test*

### 5.2 Oxidase

In this test, tetramethyl-*p*-phenylenediamine is oxidised by cytochrome oxidase to a purple compound (Figure 6.4). The test was first used to distinguish the genus *Neisseria*, but is now used in most analytical laboratories to distinguish the genus *Pseudomonas* from other Gram-negative rods. A number of variations of the test exist, which is known in the USA as the cytochrome oxidase test.

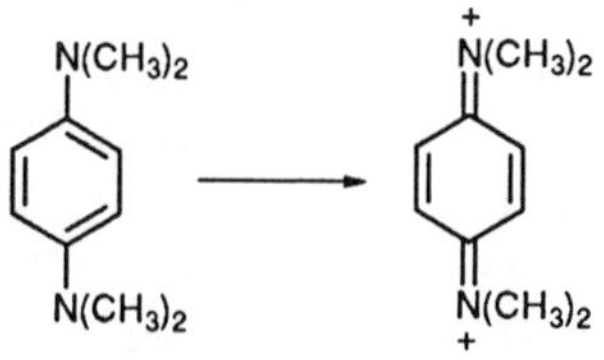

N(CH3)2 / N(CH3)2 → +N(CH3)2 / N(CH3)2+

**Figure 6.4** *Schematic representation of oxidase test*

The reagent is prepared freshly as a 1% solution and filter paper is soaked with several drops. A loopful of cells from a young culture is rubbed into the paper and a deep purple colour forms in 5–10 seconds. Any purple colour formed after about 30 seconds is regarded as negative. The reagent should not be used if purple, and, as iron catalyses the test, glass rods or platinum wires should be used. Numerous companies sell test sticks which only need to be rubbed into a colony on an agar plate. The test is an aerobic one and any cultures grown anaerobically should be exposed to air for at least 30 minutes before testing.

## 5.3 Catalase

Catalase catalyses the breakdown of hydrogen peroxide to oxygen and water. There are several methods of carrying out the test. A simple one is to place a few drops of 10% $H_2O_2$ on a slide and rub in a loopful of organisms. Evolution of gas can then be observed either with the naked eye or, if the result is uncertain, the slide may be placed under the low power of a microscope. This test is used for distinguishing aerobic from anaerobic species, especially *Staphylococcus* (catalase-positive) and *Streptococcus* (catalase negative).

It is essential that only cells are taken, as *Streptococcus* species are frequently grown on blood agar and blood is strongly catalase positive. If a culture grown on blood agar is suspected of being *Streptococcus* and is catalase positive, the test should be repeated several times.

## 5.4 Amino Acid Decarboxylases

These tests are used for diagnosis of various Enterobacteria, especially *Salmonella*. The test most widely used in this context is decarboxylation of lysine. This decarboxylation causes a rise of pH due to the formation of an amine and the indicator turns alkaline (Figure 6.5).

Commercially available tests use a combination of lysine and glucose. Organisms such as *Salmonella* which are generally lysine positive will ferment glucose initially to produce an acid reaction, and this will change to an alkaline reaction as the lysine is decarboxylated to the amine cadaverine. Lysine negative species will remain acid. The test medium should be covered with a thin layer of mineral oil to exclude air.

Other amino acids used in the identification of Enterobacteria include arginine and ornithine.

## 5.5 Amino Acid Deaminases

The enzyme phenylalanine deaminase is used for distinguishing various Enter-

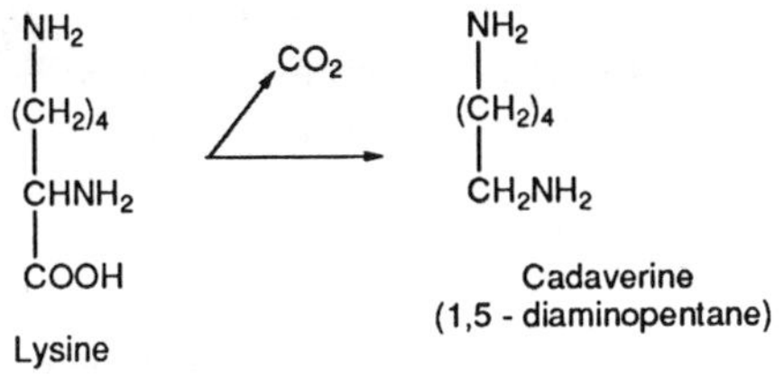

$NH_2$–$(CH_2)_4$–$CHNH_2$–$COOH$ (Lysine) $\xrightarrow{-CO_2}$ $NH_2$–$(CH_2)_4$–$CH_2NH_2$ (Cadaverine (1,5 - diaminopentane))

**Figure 6.5** *Scheme showing decarboxylation of lysine*

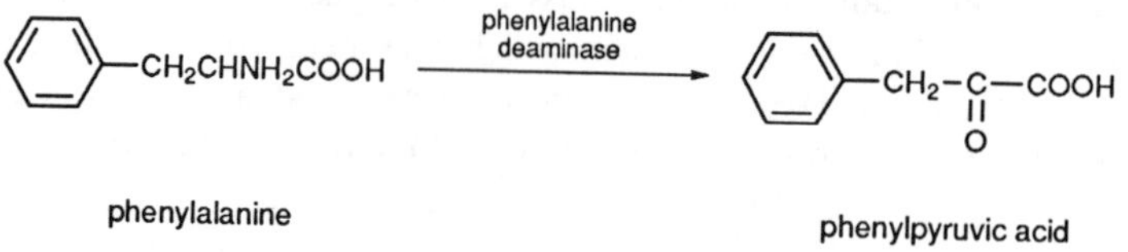

$C_6H_5$–$CH_2CHNH_2COOH$ —(phenylalanine deaminase)→ $C_6H_5$–$CH_2-C(=O)-COOH$

phenylalanine phenylpyruvic acid

**Figure 6.6** *Scheme showing the amino acid deaminase test using phenylalanine*

obacteriaceae (Figure 6.6). Phenylalanine is deaminated to phenylpyruvic acid which turns green when added to acidic ferric chloride. The test requires oxygen and must be carried out aerobically.

### 5.6 Gelatinase

This test is used for detecting bacteria which are proteolytic, *i.e.* they break down protein. Gelatin is added to the medium to solidify it. Production of the enzyme causes hydrolysis of the peptide bonds and the medium is liquified. The problem with this technique is that gelatin melts at about 27 °C, and if incubation is carried out above this temperature then the medium needs to be refrigerated for several hours before the results are noted. This test has fallen into disuse and is now rarely seen.

### 5.7 Coagulase

This enzyme coagulates plasma. The substrate is rabbit plasma which has been treated to prevent clotting. There are various methods of carrying out the test, but a simple one is to place a drop of plasma on a microscope slide and rub in a thick suspension of bacterial cells. A positive reaction is seen when the plasma clots within approximately 30 seconds. The type of plasma used is of considerable importance as sheep or horse plasma can produce different results. The test is frequently used to distinguish between pathogenic and non-pathogenic staphylococci, although the correlation between coagulase-positive cells and pathogenicity is not absolute.

### 5.8 Arginine Dihydrolase

This enzyme hydrolyses the amino acid arginine under relatively anaerobic conditions to ornithine and ammonia giving an alkaline reaction (Figure 6.7). It is used to differentiate certain types of Gram-negative aerobes, especially *Pseudomonas*. Numerous other enzyme reactions are also used but they tend to be associated with particular circumstances, *e.g.* in certain food industries.

### 5.9 Lipases

Generally these enzymes break down both simple and complex lipids but they are a poorly characterised group. The medium usually contains tributyrin (glycerol

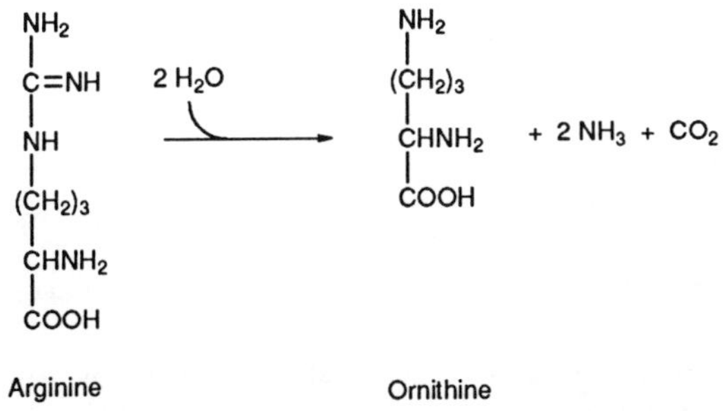

$$\underset{\text{Arginine}}{H_2N{-}C({=}NH){-}NH{-}(CH_2)_3{-}CHNH_2{-}COOH} \xrightarrow{2\,H_2O} \underset{\text{Ornithine}}{H_2N{-}(CH_2)_3{-}CHNH_2{-}COOH} + 2\,NH_3 + CO_2$$

**Figure 6.7** *Scheme showing the arginine dihydrolase test*

tributyrate), which forms an opaque suspension in the agar. Lipase positive colonies hydrolyse tributyrin producing a clearing seen as a halo round the colony.

A more complex medium containing milk or egg yolk may be used, which produces a pearly layer around colonies of clostridia and staphylococci. Free fatty acids are released and can be stained green if the plate is flooded with saturated copper sulfate.

## 5.10 Lecithinases

Lecithins are complex phospholipids and occur in serum and egg yolk. Enzyme activity breaks down the emulsion and liberates free fats so that turbidity is created. The test is used for distinguishing *Clostridium* species and may be carried out using commercially available egg yolk suspensions.

## 5.11 Proteinases

These have already been described under gelatinase. Gelatin is however a very simple protein. More complex media may be used such as Loefflers serum agar, and strongly proteolytic cultures will liquify this, eventually producing foul smelling products. Media containing milk may also be used.

## 5.12 Phosphatase

A number of substrates can be used for this enzyme. Two common ones are *p*-nitrophenyl phosphate and phenolphthalein diphosphate (Figure 6.8). In the case of *p*-nitrophenyl phosphate the *p*-nitrophenol released is bright yellow in alkaline solution and can be assayed spectrophotometrically. The test is used for the differentiation of streptococci. A variation on this test determines the phosphatase present in cow's milk as a measure of the efficiency of pasteurisation.

## 5.13 Amylases

These enzymes break down starch and are of considerable importance in industry. α-Amylase breaks down starch rapidly by attacking internal glycosidic

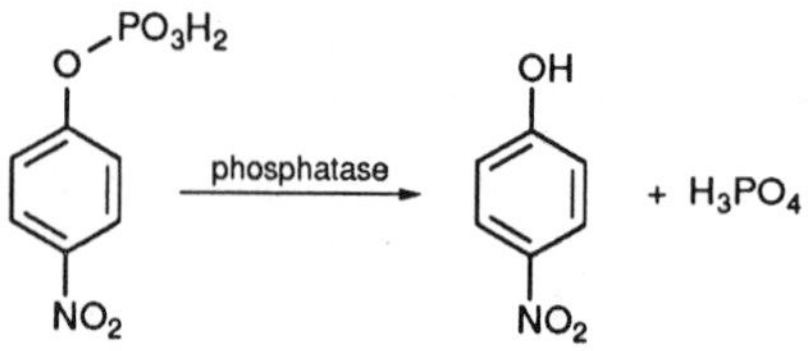

$$p\text{-}O_2NC_6H_4OPO_3H_2 \xrightarrow{\text{phosphatase}} p\text{-}O_2NC_6H_4OH + H_3PO_4$$

**Figure 6.8** *Conversion of p-nitrophenyl phosphate to p-nitrophenol using phosphatase*

bonds whereas β-amylase breaks down starch slowly by attacking the non-reducing end of the molecule. They can be demonstrated by adding iodine to the medium, a deep blue colour developing where starch has not been hydrolysed. Other depolymerases include cellulases, chitinases and pectinases. Positive tests are shown by clearing of the medium or, in the case of pectinase, liquefaction of the medium.

## 6 NITROGEN REQUIREMENTS

Micro-organisms vary widely in their ability to use different sources of nitrogen ranging from elemental nitrogen through ammonium salts to nitrate and complex amino acids. Although the ability to use these sources may be of diagnostic value, the only test used routinely is the ability to reduce nitrate to nitrite or elemental nitrogen. The ability to form nitrite from nitrate is found in virtually all the Enterobacteriaceae and the test is useful for distinguishing these from other Gram-negative organisms. The test may be carried out in either liquid medium with a Durham tube to show nitrogen formation, or in solid medium in tubes when gas production is shown by disruption of the agar.

## 7 BREAKDOWN OF LARGE MOLECULES

This has already been covered to a large extent in the section on enzymes, *e.g.* proteinases, amylases, and various lipases.

## 8 MISCELLANEOUS TESTS

### 8.1 Aesculin Breakdown

The breakdown of aesculin is used for differentiating streptococci and some anaerobes. Aesculin is the water soluble glycoside of a sterol, and on breakdown the aglycone forms an insoluble fraction which reacts with the ferric salt present in the medium to give a black/brown colour.

### 8.2 Litmus Milk

This test is rarely used due to the length of time required, but references to it may be found in older literature. Litmus is added to milk until a pale purple colour is formed. The culture is then inoculated and incubated possibly for as long a. weeks. Several different types of reaction can be seen.

(a) Acid formation. Litmus turns pink due to the production of acid from lactose.
(b) Acid clot. Litmus turns pink and the acid causes the formation of a clot of casein. After clotting, if the culture forms gas, the clot is blown apart producing a stormy clot.
(c) Curdling. Little acid is formed, the litmus stays blue but a clot is formed which slowly shrinks.
(d) Casein decomposes. The medium becomes clear and is frequently associated with an alkaline colour.
(e) Litmus is reduced and becomes colourless.
(f) Alkaline reaction caused by the citric acid in the medium being utilised.

### 8.3 Bile Solubility

Pneumococci (*Streptococcus pneumoniae)* are soluble in bile and incubation at 37 °C for 15 minutes causes clearing of suspensions. The bile activates an autocatalytic enzyme and the test distinguishes various types of streptococci.

### 8.4 Haemolysis

This refers to the effect of bacteria on agar plates containing blood. Various types of blood may be used but ox or sheep blood predominate. Many Streptococci will produce zones of complete clearing known as β-haemolysis around a colony. The term α-haemolysis is used to describe a greenish colouring of the media and this is typical of *Streptococcus viridans*. In this system the blood cells are incompletely lysed and the edge of the zone is not clearly defined. The absence of any reaction is sometimes referred to as γ-haemolysis. Horse, human or rabbit blood may also be used to differentiate various organisms as the red blood cells of different species vary widely in their sensitivity to bacterial haemolysins. Samples are incubated at 37 °C and then may be chilled to 4 °C to facilitate reading (hot/cold lysis).

## 9 REFERENCES AND FURTHER READING

1. S. Bascomb, 'Enzyme Tests in Bacterial Identification', Methods in Microbiology, Vol. 19, ed. R.R. Colwell and R. Grigorova, Ch. 3, pp. 105–160, Academic Press, London and New York 1987.
2. A.J. Holding and J.G. Collee, 'Routine Biochemical Tests', Methods in Microbiology, Vol. 6A, ed. J.R. Norris and D.W. Ribbons, Ch. 1, pp. 1–32, Academic Press, London and New York, 1971.
3. K. Kersters and J. de Ley, 'Enzymic Tests with Resting Cells and Cell-Free Extracts', reference 2, Ch. 2, pp. 33–52.
4. B. Lanyi 'Classical and Rapid Identification Methods for Medically Important Bacteria'. reference 1, Ch. 1, pp. 1–67.

# Practical Identification of Bacteria

It should be realised that the identification of bacteria is heavily biased in favour of species of medical and industrial importance. Identification at a practical level in the laboratory involves a number of categories of tests. These are (a) morphological, (b) cultural, (c) biochemical and physiological, (d) serological, and (e) phage typing. Serological tests and phage typing are not usually carried out in an analytical laboratory, as these are mainly epidemiological techniques and would be done in a Public Health Laboratory.

**In all cases it is absolutely essential that the tests are carried out on pure cultures. Tests carried out on mixed cultures are of no significance whatsoever.**

Initially the sample should be streaked out onto a plate of non-selective medium and grown for 24/48 hours. Once the organism has grown, a single well isolated typical colony should be chosen and used for all tests. It is bad practice to choose several similar looking colonies and use these, although it has to be accepted that there are occasions when this is unavoidable. If there is difficulty in choosing a single colony, it may be easier if the plate is placed under the low power of a microscope. If there is a suspicion that the culture is not pure, or there are serious difficulties in picking out single colonies, then a colony should be streaked out onto a second plate of non-selective media. It is also advisable to use young colonies of bacteria, because if fungal contaminants are present and commence sporing, bacteria become significantly more difficult to purify.

Inoculation of the sample onto selective media is bad practice, as this does not produce a pure culture but only suppresses the growth of unwanted species. Although experience suggests that the same problem is usually caused by the same or similar organisms, occasionally some unusual organism may be present. This may survive the selective medium, and start to grow once the organism is subcultured into various test media.

Once the bacteria have grown on the non-selective medium, cultures should be examined for their colonial morphology, and also to determine whether the organism has had any effect on the medium. Generally cultural characteristics on simple media are too variable to be of diagnostic value and, although great emphasis was put on this in older literature, it is now considered to be of little use and the glossary of terms used to describe colonial morphology is generally considered redundant. However, a few groups of bacteria produce characteristic morphology which may allow the experienced microbiologist to make an educated guess.

It must be emphasised that common organisms are common, and when a sample from some well documented source is examined, it is most unlikely to find some unusual genus or species. If this happens, it is highly probable that the culture is not pure or that some test has been misread. In this case, it is advisable to reisolate the organism from the sample and repeat all the tests. It is therefore obvious that the samples should not be discarded until all the tests have been completed, and a suitable answer has been obtained. Occasionally, however, samples do contain an organism so unusual that it cannot be identified from the standard tests. It should also be emphasised that if a sample has been damaged microbiologically or if such damage is imminent, then the number of micro-organisms present will be very high.

Identification on a practical basis can follow several routes. These include the traditional (hierarchical) methods which place heavy reliance on staining and morphology in the initial phase and only move to biochemical testing at a later stage, and the more recent systems based on Adansonian classification such as the Bio-Merieux API systems which use biochemical methodology from the outset. The major differences between the methods is that the hierarchical method gives different weightings to various criteria and uses them in a specific order. In many cases, the order may change or later criteria may be omitted depending on the results of earlier tests.

In numerical classification and identification, all the tests are carried out simultaneously, and the same weighting is given to each character. One major problem is that not all members of the species or genus may give the same result. It is not unusual to find that only 80 or 90% of the organisms tested give a positive result. This can cause considerable problems in programming computers and lead to considerable confusion.

When the hierarchical method is used, the main groups of bacteria can sometimes be distinguished by their cultural appearance on solid media, microscopic examination of their shape (morphology), and by their staining reactions. The staining reactions follow a sequence in which the Gram stain is used first, and the stains used subsequently depend on the result obtained in the Gram stain. Further important criteria in this methodology are motility, and the presence or absence of spores.

It is only when these criteria have been determined that biochemical tests are applied. The majority of the tests used are based on the ability to grow on specific sugars and the presence or absence of specific enzymes. Further tests may be applied; these include antigenic tests which make use of the bacteria as an antigen and test it against specific antibodies, and also the use of phage typing in which the bacteria is challenged with a variety of phages (bacterial viruses).

The use of 'rapid' systems, such as the API system, relies on a series of biochemical tests carried out simultaneously. The results of these tests are pooled and compared with sets of 'ideal' results. This comparison can either be carried out manually or more usually on a computerised data base to obtain a nearest match. Essentially, the system is a 'nearest neighbour' analysis carried out in $n$-dimensional space, where $n$ is the number of parameters measured. Many of the tests used are considered in the chapter on biochemical testing.

Whilst these tests are rapid, they do have certain failings in that they do not detect the contaminated culture as rapidly as the traditional hierarchical system, in which the contaminant is usually picked out by aberrant morphology or staining at an early stage. There is also the suspicion that the increase in speed has been attained at the expense of a better understanding of identification methodology. This problem is usually encountered when an aberrant culture is obtained which does not fit the data base.

There are several methods of dealing with the practical problems of identification. Cowan and Steel[1] suggest three alternative approaches to dealing with unknown organisms. These are:

(a) A blunderbuss approach in which every test is done and the results are compared with Bergey.[2]
(b) A probability approach which involves an informed assessment of the micro-organisms likely to be present in a given situation. For example, in water suspected of being contaminated by sewage, one is looking for thermotolerant coliforms, faecal streptococci and *Clostridium perfringens*. Tests would be carried out to indicate the presence or absence of these groups. If this approach is used, then it is necessary to be aware that something unusual may be present.
(c) The progressive or step by step approach, which is the one most commonly used. This is obviously a hierarchical approach. The first stage involves the determination of several fundamental characters. These include the shape, Gram stain, motility, and several tests or further stains depending on the result of the Gram stain. Once these results are known, which may take 24 hours if any simple growth tests have been included, then a second stage series of tests is carried out. The tests included in the second stage will vary depending on the results obtained in the first stage. It may be possible to identify the organism from the second stage tests, or it may be necessary to carry out third stage tests to determine the species present. First stage tests are very similar for all groups, whilst the second and third stage tests vary depending on the group under consideration.

The best tests are those which are quick, simple to perform, stable and not subject to operator error. Characters such as pigmentation are too variable in the majority of groups to be of much use in identification. Whilst they may be of use in a few genera, pigmentation may be influenced significantly by the medium, light or dark, and temperature. It should also be noted that a character which may be of use in identifying one genus to species level, may be so variable in another genus that it is of no diagnostic use whatsoever.

Antibiotic resistance is also not a stable character in most cases and is used mainly in medicine. Resistance to metronidazole may be of use, as only obligate anaerobes are sensitive to this compound. Facultative anaerobes and aerobes are not sensitive and this property may be used to distinguish these groups.

Similarly, when considering ease of use, motility is a very useful diagnostic test and is easy to perform. However, the position of the flagella, which is also of diagnostic value, is a very difficult test to carry out and is rarely done on a routine basis.

Cowan and Steel[1] give details of first, second and third stage tests for a large number of medically important species, and also a series of mini definitions for various genera.

## 1 FIRST STAGE TESTS

A sample of cells should be taken from a single colony and stained using the Gram stain. This is the most important differential stain in microbiology. The Gram reaction and the morphology of the cells may be examined at the same time. The morphology of cells (*i.e.* the shape, size, and appearance), taken with their Gram reaction, and the source from which they were isolated, may enable the experienced observer to identify a culture to its probable generic level, but identification beyond this point requires other methods.

When a culture is examined microscopically for its morphological characteristics a number of points need to be noted. The initial one is the appearance of the cells, whether they are cocci or rods, and whether the cells are arranged singly, in filaments, in groups, or in chains. In a few groups of bacteria, it may be difficult to decide whether the cells are cocci or rods. This problem may be related to the stage of the growth cycle from which the cells were taken.

The Gram stain (Chapter 4) should be carried out on fresh cultures grown overnight, as old cultures frequently give a Gram-negative response. This stain indicates differences in cell wall chemistry. The crystal violet and iodine form a complex which cannot be removed from Gram-positive cells using ethanol, and cells thus appear purple. Gram-negative cells, however, are decolourised and when counterstained appear pink/red. When handling unknown cultures, it is sound practice to stain cells whose Gram reaction is known at the same time.

If the results show that a Gram-positive organism is present, then the first stage tests shown in Table 1 should be carried out. If the organism is Gram-negative, then all the tests shown in Table 2 should be carried out.

In the spore stain (Chapter 4), cells are being examined for the presence of resistant spores. The test works on the basis that the dye, malachite green, is driven into the spores by the heating process and cannot be removed by washing with distilled water. Washing does remove the dye from the vegetative portion of the cell, which can then be counter-stained with safranin. Spores therefore stain green, and the vegetative portion of the cell stains red.

If spores are present, then the organism is either from the genus *Bacillus* or *Clostridium*. These can easily be distinguished as *Bacillus* is aerobic and *Clostridium* is anaerobic. *Bacillus* sometimes loses the ability to sporulate. This may be permanent, but can frequently be reversed by changing medium or by growing

**Table 1** *Gram positive bacteria (first stage tests)*

| | *Shape* | *Acid fast* | *Spores* | *Motile* | *Grows in air* | *Grows anaerobically* | *Catalase* | *Oxidase* | *Acid from glucose* | *O/F*[c] |
|---|---|---|---|---|---|---|---|---|---|---|
| *Micrococcus* | C | – | – | – | + | – | + | + | V | O/- |
| *Staphylococcus* | C | – | – | – | + | + | + | – | + | F |
| *Streptococcus* | C | – | – | – | + | + | – | – | + | F |
| *Enterococcus* | C | – | – | V | + | + | – | – | + | F |
| *Lactococcus* | C | – | – | – | + | + | – | – | + | F |
| *Leuconostoc* | C | – | – | – | + | + | – | – | + | F |
| *Corynebacterium* | R | – | – | – | + | + | + | – | + | F |
| *Listeria* | R | – | – | + | + | + | + | – | + | F |
| *Lactobacillus* | R | – | – | – | + | + | – | – | + | F |
| *Propionibacterium* | R | – | – | – | – | + | + | ? | + | F |
| *Clostridium* | R | – | +[a] | V | V | + | – | ? | V | F/- |
| *Bacillus* | R | – | +[a] | V | + | V | + | V | V | O/F/- |
| *Nocardia* | R | S/+ | – | – | + | – | + | – | + | O |
| *Mycobacterium* | R | + | – | – | + | ? | + | – | + | O |
| *Streptomyces* | M | – | +[b] | – | + | ? | + | ? | ? | ? |

C = coccus, R = rod, M = mycelial.
+ = 85–100% positive, – = 0–15% positive, S/+ = slight, V = variable.
[a]endospores, [b]exospores, [c]Hugh and Leifson's oxidation/fermentation tests.
Modified with permission from 'Cowan and Steel's Manual for the Identification of Medical Bacteria', 3rd Edn., ed. G.I. Barrow and R.K.A. Feltham, Cambridge University Press, 1993.

**Table 2** *Gram negative bacteria (first stage tests)*

| | *Shape* | *Motile* | *Grows in air* | *Grows anaerobically* | *Catalase* | *Oxidase glucose* | *Acid from* | *O/F*[c] |
|---|---|---|---|---|---|---|---|---|
| *Brucella* | R | – | + | – | + | + | – | – |
| *Alcaligenes* | R | + | + | – | + | + | – | – |
| *Pseudomonas* | R | + | + | – | + | + | V | O/- |
| *Achromobacter* | R | + | + | – | + | + | + | O |
| *Aeromonas* | R | + | – | + | + | + | + | F |
| *Enterobacteria*[d] | R | V | + | + | + | – | + | F |
| *Vibrio* | R | + | + | + | + | + | + | F |
| *Legionella* | R | + | + | – | + | V | – | – |
| *Campylobacter* | a | + | b | V | V | S/+ | – | – |

R = rod.
+ = 85–100% positive, – = 0–15% positive, S/+ = slight, V = variable.
[a] Helical or curved cells, [b] Grow in 3–10% oxygen, [c]Hugh and Leifsons oxidation/fermentation test, [d] Includes the following genera mentioned in this book, *Enterobacter*, *Erwinia*, *Escherichia*, *Klebsiella*, *Proteus*, *Salmonella*, *Shigella*. These can be separated utilising various tests and media given in the text.
Modified with permission from 'Cowan and Steel's Manual for the Identification of Medical Bacteria', 3rd Edn., ed. G.I. Barrow and R.K.A. Feltham, Cambridge University Press, 1993.

the cells on a medium containing soil extract. The position and size of the spores should be noted as these may be of diagnostic value, *i.e.* are they central, subterminal or terminal, and are they the same diameter as the cell or do they cause the cell to bulge?

It is not necessary to carry out an acid fast stain (Chapter 4) on Gram-positive rods giving a positive spore stain, but this test should be carried out on Gram-positive rods which cannot be demonstrated to contain spores. This stain is intended to differentiate the genus *Mycobacterium* and some *Nocardia* spp. on the basis of the lipid material present in the outer layers of the cell.

Several different protocols are found for each of these stains, and it is best if laboratories standardise on one method. A number of companies market equipment which allows the simultaneous staining of multiple samples. These can obviously save considerable time. A wide variety of other stains are found in the literature, but those given above are the ones of greatest diagnostic value.

Motility must also be checked. This may be done in one of several ways, for example by direct observation using the hanging drop method, by flagellar staining, or by a cultural method. If the first two methods are attempted, then it is essential that a young liquid culture is used. The temperature of incubation is also important as a number of organisms are motile at 30 °C but not at 37 °C, even though this may be the optimum temperature for growth. The hanging drop method and flagellar staining both require some experience to produce reliable results. The cultural method is the most reliable in inexperienced hands. This involves the preparation of a bottle of a sloppy agar medium with a glass tube in the middle which is open at both ends (Craigie tube). The central tube is inoculated with bacteria using a straight wire, incubated overnight, and examined for growth outside the tube. If growth is confined to the centre of the tube then the organism is non-motile, but if growth is found throughout the medium, then the cells must have been motile.

Both the oxidase and catalase tests are very rapid, producing results in 2–3 minutes. Details may be found in the chapter on biochemical methods.

Cultural tests carried out as primary tests include the ability to produce acid from glucose, and the Hugh and Leifson oxidation/fermentation (O/F) test. Details are found in the chapter on biochemical methods.

## 2 SECOND AND THIRD STAGE TESTS

The division of tests into second and third stage tests depends on the group under investigation. They usually include the breakdown of a variety of sugars and sugar alcohols using a liquid medium containing an inverted Durham tube to determine gas production. Other commonly used tests are the indole, Voges–Proskauer, $H_2S$ production and various enzyme tests described in the chapter on biochemical testing. A detailed breakdown of second and third stage tests for many groups is found in Cowan and Steel.[1]

Simple flow charts are given in Figures. 7.1, 7.3, and 7.6 (see later for Figures 7.2 and 7.6) for the identification of a number of major genera, but these should be used in conjunction with the description of each genus in the text.

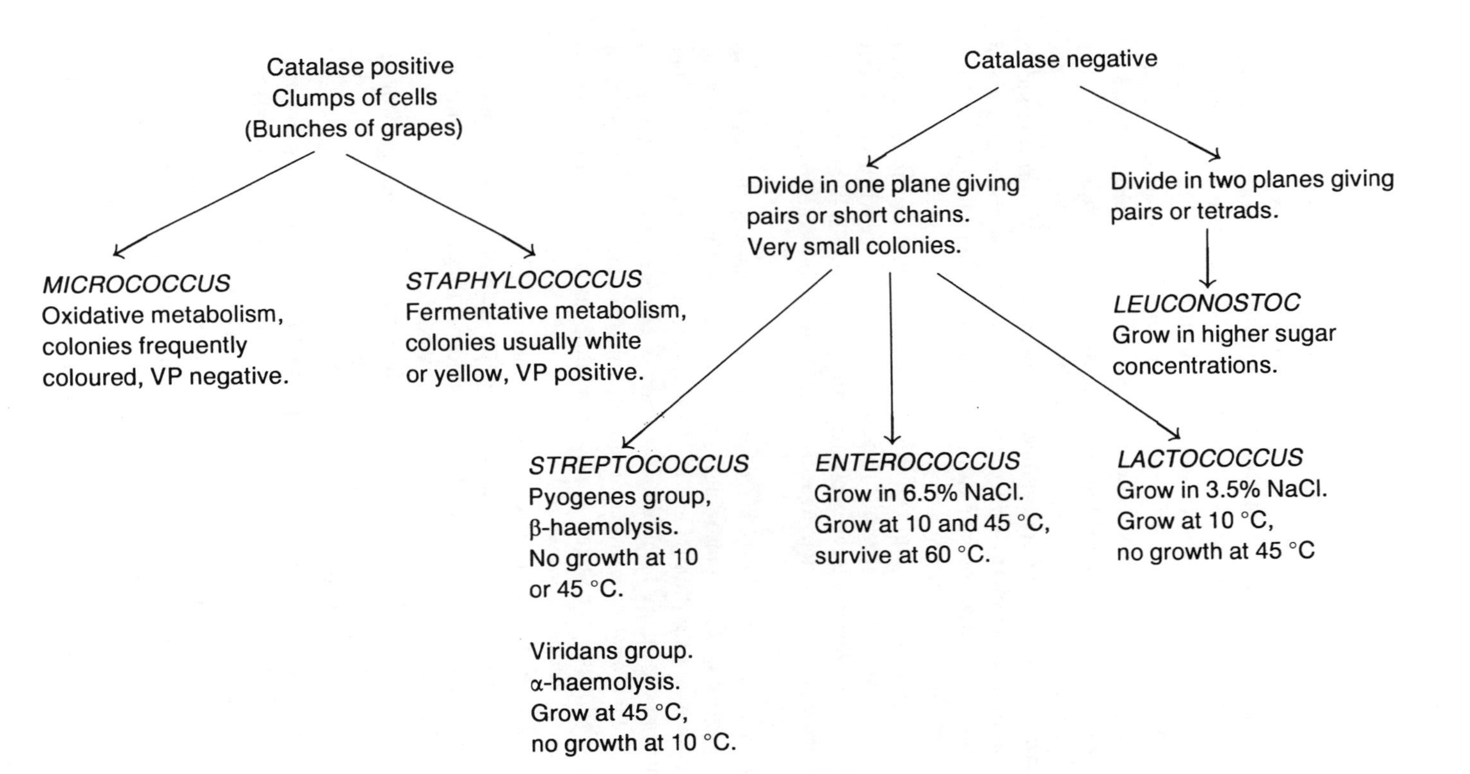


**Figure 7.1** *Gram-positive cocci*

## 3 GRAM-POSITIVE BACTERIA

### 3.1 Gram-positive Cocci

Gram-positive cocci are members of the genera having coccoid morphology shown in Table 1. *Micrococcus*, *Staphylococcus*, *Streptococcus*, and *Enterococcus* are all of importance in food and water. A tentative differentiation may be made between *Streptococcus*, and *Enterococcus* and the other groups on the basis of their microscopic appearance (Figure 7.2). Streptococci are usually seen as short chains, whereas the other cocci are usually clumped. Streptococci can also be distinguished by the fact that they are catalase negative. Streptococci also tend to produce smaller colonies than other cocci when grown on agar under aerobic conditions.

*3.1.1 Micrococcus.* These are of importance as they are common forms found in air and dust, and large numbers of micrococci in a sample may be indicative of poor hygiene and a potential spoilage situation. They may be distinguished from the staphylococci, to which they are closely related, by their oxidative reactions when grown on carbohydrates, and can therefore be identified by the Hugh and Leifson test. They may also be distinguished by the Voges–Proskauer test (Chapter 6), in which *Staphylococcus* is positive and *Micrococcus* is negative. They grow well on media such as nutrient agar frequently producing coloured colonies.

*3.1.2 Staphylococcus.* Staphylococci are of importance because of their ability to produce toxins causing food poisoning. They may be distinguished by their growth on media high in sodium chloride. This suppresses growth of most other species which do not have this level of halotolerance. Two species of *Staphylococcus* are usually identified. *S. aureus* grows strongly on a mannitol/salt medium, which contains salt (7.5%) and the sugar alcohol mannitol, usually forming yellow/golden colonies. *S. epidermidis* grows rather less strongly on this medium usually producing red colonies. Other species are usually inhibited, and the medium can be used for the direct counting of staphylococci from food and milk. Yellow colonies growing on mannitol/salt medium should be treated as presumptive coagulase positive.

These two species of staphylococci may be differentiated by their growth on Baird–Parker medium. Several versions of this medium are found. The first contains egg yolk and tellurite, the egg yolk making the medium yellow and opaque. *S. aureus* grows, reducing the tellurite to form black, shiny colonies surrounded by a halo formed by proteolysis of the egg yolk. The second form of Baird-Parker medium replaces the egg yolk with blood, using this to give a direct result for the coagulase test. *S. aureus* produces grey/black colonies with an opaque halo around them caused by blood coagulation. Whilst there is not a total correlation between coagulase production and the pathogenicity of *S. aureus*, the relationship is good. Alternatively, this reaction may be distinguished using the method for coagulase given in the chapter on Biochemical Testing. Baird–Parker medium may also be used for the direct counting of *S. aureus* in food.

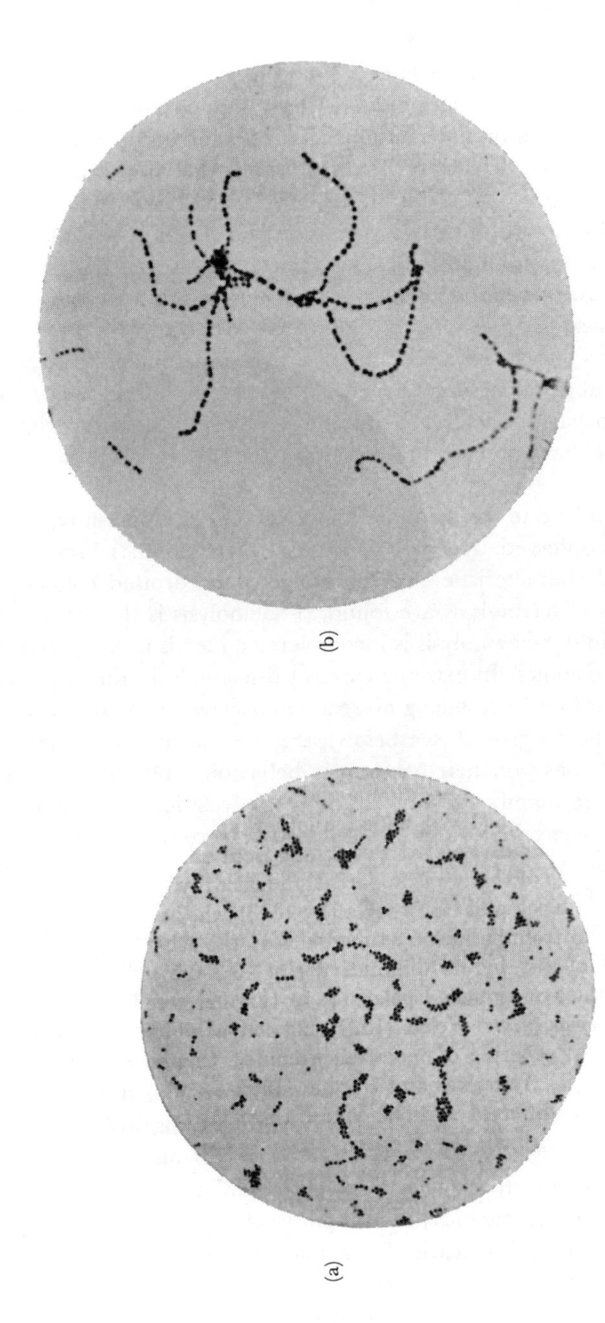


**Figure 7.2** (a) *Staphylococcus aureus* (× 1000) *showing short chains and clumps of cells.* (b) *Streptococcus pyogenes* (× 1000) *showing long chains of cells*
(Reproduced by permission from 'Topley and Wilson's Principles of Bacteriology and Immunity', 4th Edn., ed. G.S. Wilson and A.A. Miles, Edward Arnold, 1957)

*3.1.3 Streptococcus.* The streptococci are another important group whose presence is frequently associated with problems. They are commonly found in the throat, a number of dairy products, and also in water supplies contaminated with sewage or faecal material (see later).

Streptococci may be distinguished from staphylococci by the size of their colonies when grown on nutrient blood agar, streptococcal colonies usually being significantly smaller than staphylococcal ones. The streptococci are microaerophilic and only grow poorly in air as they lack a cytochrome system. One test which does distinguish these two groups unequivocally is the catalase test, details of which are given in Chapter 6. It is important, however, that if the streptococci have been grown on blood agar, no blood is picked up with the cells, as blood is strongly catalase-positive.

Streptococci may be isolated from materials contaminated with a wide range of bacteria by subculturing onto a medium containing sodium azide which inhibits most Gram-negative bacteria. Sodium azide acts on the cytochrome system and as streptococci have no cytochrome system, they are immune to the effects of this chemical.

If blood is added to the medium then a number of different types of haemolysis may also be observed. Haemolysis occurs when blood is broken down by the bacteria, and characteristic clearing of the blood around colonies takes place. Several types of haemolysis are found, β-haemolysis is the complete clearing of the blood, whilst α-haemolysis is incomplete and leads to the presence of a green halo around colonies. In extreme cases of β-haemolysis, the entire plate may be cleared. A third type producing no reaction is known as γ-haemolysis. Before the use of Lancefield's groups (see below), the streptococci were divided into three groups on the basis of their haemolytic behaviour. The pyogenic (pus forming) streptococci are mainly β-haemolytic, but the relationship is not absolute. The second group were the viridans group producing α-haemolysis. The third group were those not carrying out haemolysis.

Azide blood agar is widely used for the isolation of streptococci from a number of sources. Azide may become explosive under certain conditions and the manufacturer's instructions should be followed for disposal of media.

The present classification of the streptococci is based on the presence of various complex carbohydrates known as Lancefields antigens. This allows them to be split into four groups. These are the pyogenic streptococci, the viridans group, the enterococci and the milk or lactic streptococci. Attempts to match these groups with biochemical characteristics run into serious problems.

The pyogenic streptococci include a number of pathogenic species found in raw milk, especially that taken from cattle suffering from mastitis. The most important of the pathogenic species is *Streptococcus pyogenes*, the causative organism of sore throats, tonsilitis, scarlet fever, and rheumatic fever. Identification of the pyogenic streptococci is best carried out on the basis of serology in a Public Health Laboratory. However, a tentative identification may be carried out on the basis of their haemolytic reaction, and the inability of this group to grow at 10 or 45 °C.

The viridans group includes *S. thermophilus*, which is found in fermented milk

**Table 3** Characteristics of *Lactococcus* and *Streptococcus*

| | *Lactococcus* | *Streptococcus* |
|---|---|---|
| Haemolysis | -/α | mainly β |
| Phosphatase | -ve | mainly +ve |
| Grows in 4% NaCl | most | very few |
| Voges–Proskauer | +ve | mainly -ve |
| Aesculin hydrolysis | +ve | variable |
| Ribose fermentation | +ve | variable |
| Lancefield antigens | N | variable but not N |

drinks, yoghurts, and certain cheeses. These may be distinguished by their α-haemolysis and their ability to grow at 45 °C but not at 10 °C. The L-S differential medium of Oxoid may be used to count *S. thermophilus* in the presence of lactobacilli in, for example, yoghurt.

The third group, the enterococci or faecal streptococci, are extremely important in the microbiology of water, where their presence is indicative of faecal contamination. The isolation and identification of this group is considered in detail (Chapter 9). The enterococci have been split relatively recently from the streptococci, but DNA studies show that they are a separate group, and the organism *Streptococcus faecalis* is now known as *Enterococcus faecalis*. It is, however, still referred to as *S. faecalis* in many books on water microbiology. *Enterococcus* spp. can be separated from *Streptococcus* spp. by their ability to grow in media containing 6.5% sodium chloride, and their growth at 10 °C and 45 °C. They also survive heating to 60 °C for 30 minutes.

A medium widely used for detecting and counting enterococci is Slanetz and Bartley's medium. The medium contains azide and the dye tetrazolium chloride. When incubated at 44 °C, the enterococci form red/maroon colonies, and the medium can be used for the isolation of enterococci from both food and water.

The lactic streptococci, the most important of which was originally known as *S. lactis*, have been transferred to the genus *Lactococcus*. They are of considerable importance in the milk and cheese industries. Differentiation of this group from the genus *Streptococcus* on the basis of biochemical tests is difficult, but a comparison of characteristics is shown in Table 3. They can tolerate up to approximately 3% sodium chloride, and will grow at 10 °C but not 45 °C.

*3.1.4 Leuconostoc.* This genus contains cocci carrying out a heterofermentative type of metabolism producing a mixture of lactic acid, ethanol, acetic acid, and carbon dioxide. They are used in the dairy industry for their ability to produce diacetyl, an important flavouring compound in butter, buttermilk, and certain cheeses. They do, however, also cause spoilage in foods high in sucrose, as they are able to grow in concentrations of up to 60% sugars producing a slime of dextrans.

### 3.2 Gram-positive Rods

A number of Gram-positive rods (Figure 7.3) are of importance as certain genera are used in industry, whilst others are found in spoilage and food poisoning situations.

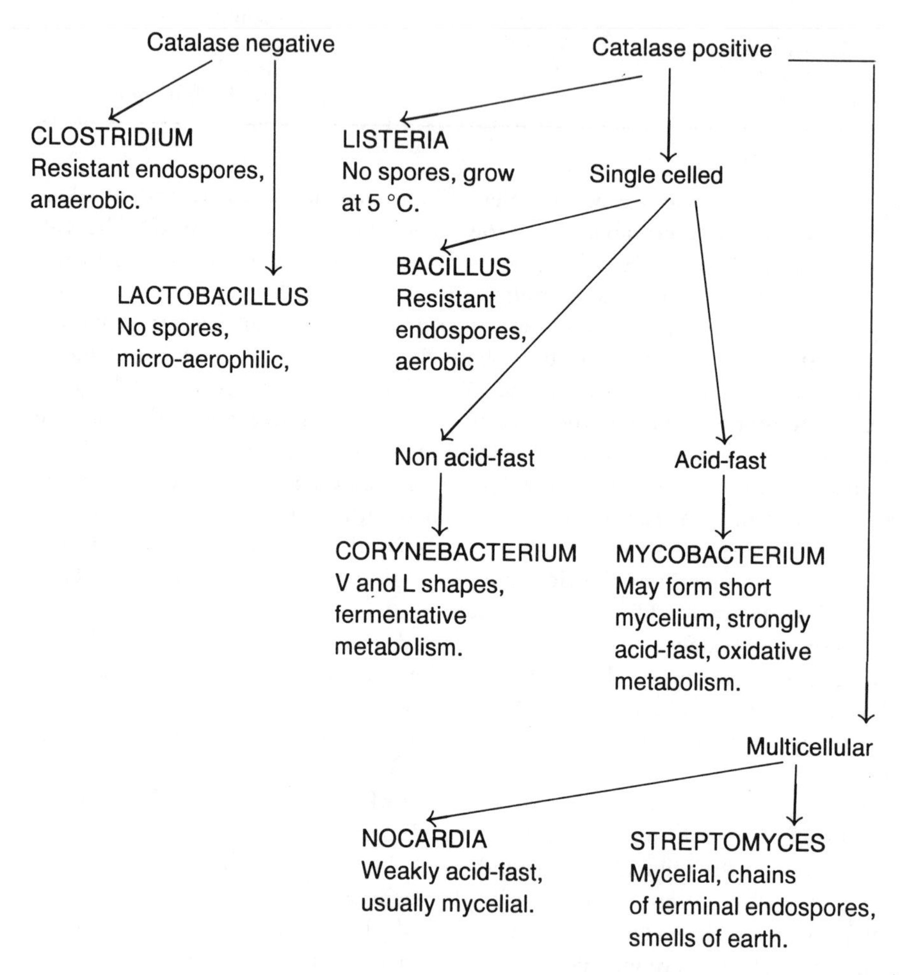


**Figure 7.3** *Gram-positive rods and coccobacilli*

*3.2.1 Corynebacterium.* This group is a very heterogeneous one containing a number of pathogenic organisms. The Gram stain should always be carried out on young cultures and, although the organism is defined as a Gram-positive rod, on microscopical examination it usually appears as a club shaped organism which may be slightly curved. The rods also tend to be arranged in a manner sometimes referred to as Chinese lettering. The differentiation of the corynebacteria is based on chemical characteristics which are not easy to determine in a routine laboratory. The parasitic and pathogenic species show a common feature with the mycobacteria in that they possess mycolic acids, long chain fatty acids with an α-branching group and a β-hydroxy group.

The most important pathogen in the group is *C. diphtheriae*, the causative organism of diphtheria. This can be divided into three sub-groups, gravis, intermedius, and mitis, which can be differentiated by their growth behaviour in broth and their growth on solid media containing blood and tellurite. A significant portion of the population are carriers of corynebacteria in their throats, *i.e.* they have the organism present in their bodies but do not present any clinical symptoms. The Tinsdale medium of Oxoid[3] or the Mueller tellurite medium of Difco[4] may be used for growth and identification.

*3.2.2 Listeria.* These are small rods which are motile when grown at 20–25 °C, but usually non-motile at 37 °C, which enables them to be differentiated from *Corynebacterium*. The most important species is *L. monocytogenes*. They are commonly found in soil and vegetation, and are the cause of a number of pathogenic conditions in humans, including spontaneous abortions. There is a relatively high rate of mortality, especially amongst immunocompromised patients. They are found in a number of pre-cooked and chilled foods, including cold meats, non-pasteurised cheeses, pre-packed salads, and poultry. This causes serious problems as they are able to survive and grow at temperatures of 1–5 °C. They can be isolated selectively by growth in a refrigerator, but growth is so slow that they are usually grown at 30 °C, on a selective enrichment medium, containing an antibiotic such as nalidixic acid to suppress the growth of other organisms. A number of commercial media are available, many of which include aesculin as a marker, enabling *Listeria* to be identified by the black zone around the colony.

*3.2.3 Lactobacillus.* *Lactobacillus* is not regarded as a pathogenic group although some are parasitic in the mouth contributing towards dental caries. They are an extremely important genus industrially and are also involved in numerous spoilage situations. Industrially, they play an important role in the souring of milk to form yoghurt and cheese, and they are also involved in the formation of acid during the production of pickles, sauerkraut, and silage. They have a very high tolerance to acid and, by producing large quantities of lactic acid, they are able to eliminate many other species from mixed populations. They are either microaerophilic or anaerobic. They may be grown on media such as tomato juice agar or MRS medium at a pH of 6.0, which is lower than most bacteria prefer. If heavy contamination by other species is expected, the pH may be lowered to 5.0–5.2 which eliminates most other organisms.

*Lactobacillus* is also an important group of spoilage organisms, especially in foodstuffs high in sugars. In addition, they are responsible for the production of mousy flavours in wine, and may also cause spoilage in meat.

They have very fastidious requirements for the B group of vitamins and may be sub-divided on their nutritional requirements. They may be grown and counted on the commercially available MRS medium, preferably under microaerophilic conditions. These nutritional requirements may be utilised for the bioassay of certain vitamins such as cyanocobalamin (see Chapter 11).

*3.2.4 Propionibacterium.* This genus is also of importance in cheese making where it develops as a member of the secondary microflora, and plays a role in ripening. As the name suggests, it forms propionic acid from sugars and also large amounts of carbon dioxide which contribute to the formation of holes. Some pigmented strains may cause spoilage by colouring the cheese.

*3.2.5 Bacillus.* Two genera of bacteria produce endospores (inside the cell); these are *Bacillus* and *Clostridium.* The spores are highly resistant and, in many cases, capable of surviving inadequate cooking or sterilisation procedures, causing spoilage or food poisoning at a later date. However, the existence of spores makes it relatively easy to detect the presence of these bacteria, and also distinguishes them from the genera *Lactobacillus* and *Listeria.* The two groups may be distinguished by the fact that *Bacillus* is aerobic and *Clostridium* is anaerobic.

The genus *Bacillus* is an extremely heterogeneous one, and organisms are allocated to this genus primarily on the basis that they are aerobic spore forming rods. The majority of members are Gram-positive, but it is essential that the Gram stain is always carried out on young cultures, as there is a marked tendency for old cultures to become Gram-negative. A few strains may be Gram-negative which can cause considerable confusion. In addition, there are strains which do not form spores, or which lose the ability to sporulate readily. In the latter case, the ability to form spores may be regained by growing the organism in a medium containing soil extract. The position and shape of the spores is of diagnostic importance, although it may take considerable practice to distinguish these. Other variable features include motility, the ability to grow anaerobically, the oxidase reaction, and the pathway of carbohydrate breakdown. Most of the important species are Gram-positive, aerobic, and generally form spores.

Medically, the most important member of the group is *B. anthracis,* the causative organism of anthrax. Whilst this is not common in the UK, there is always the possibility that it may appear in material of natural origin, and the laboratory must always be aware of this possibility. The danger of this organism is that it is highly invasive, the onset of disease is extremely rapid and the disease is frequently fatal. A number of authors regard *B. anthracis* as a pathogenic form of *B. cereus* and it may be distinguished from *B. cereus* by a number of tests. These include motility, capsule formation, behaviour on blood agar, and phage lysis tests. In the interests of safety, if there is any suspicion that *B. anthracis* is present, then the raw material should be sealed and referred immediately to a pathology laboratory accompanied by a full history. All samples and plates should be destroyed immediately by autoclaving.

The most important member of the genus *Bacillus* in terms of food poisoning is *B. cereus* (Figure 7.4), which is found frequently in samples both as an indicator and as a pathogen. This is the most likely species to be found in an analytical laboratory. It may be isolated by growth on a *B. cereus* selective medium containing mannitol, egg yolk emulsion, the antibiotic Polymyxin B, and Bromothymol Blue. *B. cereus* does not metabolise mannitol but does hydrolyse lecithin, a complex lipid present in egg yolk, to form a precipitate. Colonies of *B. cereus* may be recognised by their blue colour, showing their inability to ferment

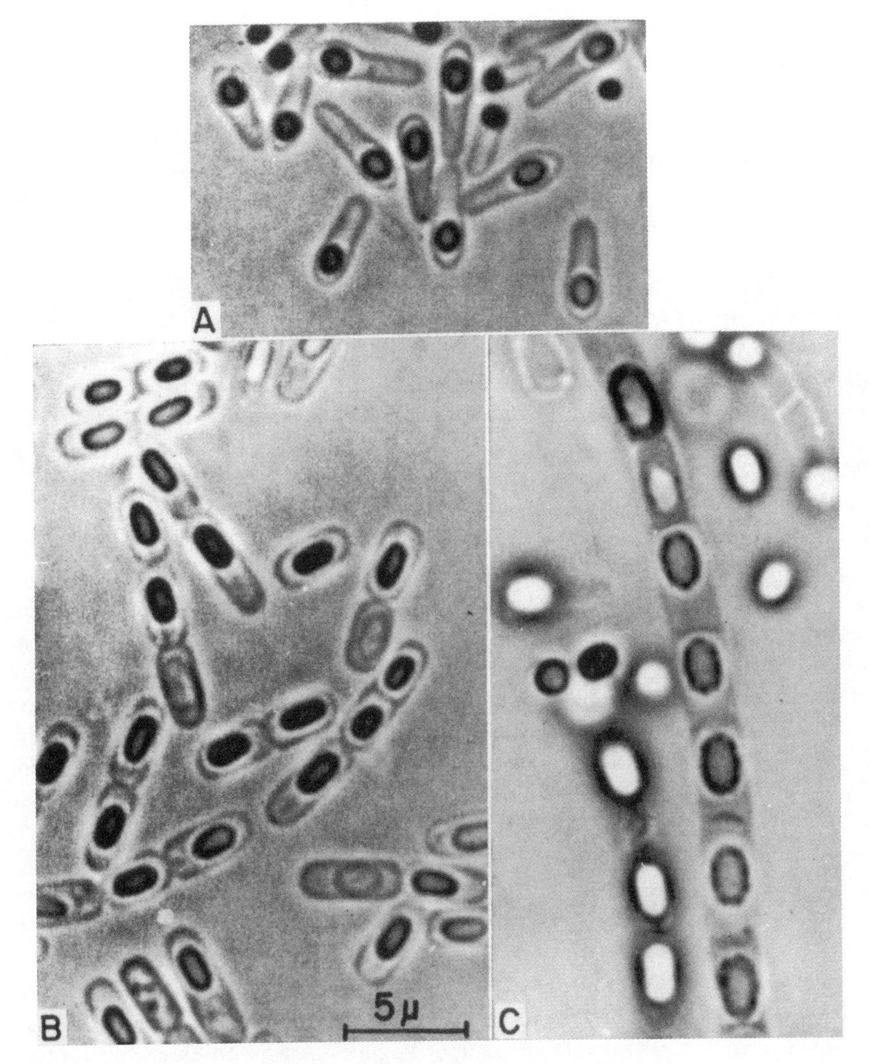


**Figure 7.4** *Bacillus cereus showing chains of cells and sporulation*
(Reproduced by permission from I.C. Gunsalus and R.Y. Stanier, 'The Bacteria', Volume 1, Academic Press, London, 1960)

mannitol, and the production of an opaque zone around the colonies. The medium does allow the growth of some other species, but they do not produce the typical colour reaction. Small numbers of *B. cereus* are usually harmless, and the role of the organism in food poisoning should only be suspected if numbers are of the order of $10^8\,g^{-1}$. The most frequent cause of food poisoning by this organism is from cooked rice which has been kept under poor storage conditions, thus allowing the formation of an enterotoxin.

Other species which may be found include *B. subtilis*, *B. mycoides*, and *B. megaterium*. Three important industrial species are *B. stearothermophilus*, a thermophilic organism used for checking the efficiency of autoclaves, *B. thuringiensis*, the spores of which are used as an insect pathogen, and *B. amyloliquefaciens*, which produces amylase used for converting starch to soluble sugars.

An important medium for the growth of *B. stearothermophilus* is Spore Strip Broth. Strips containing spores of *B. stearothermophilus* are frequently placed in autoclaves to determine their efficiency. When removed after autoclaving, these strips are placed in the broth and incubated at 55 °C for 7 days. Growth indicates that the autoclaving process was deficient.

*3.2.6 Clostridium.* Enteropathogenic clostridia and *Clostridium botulinum* are another group of Gram-positive rods associated with food poisoning. A mild form of food poisoning may be caused by some strains of *C. perfringens* type A. The division into various types (A-E) is on the basis of the toxin pattern produced. The usual form is type A2 which are non-haemolytic on horse blood agar, and form heat resistant spores, although type A1 (classical) are also found. These are β-haemolytic and have relatively heat sensitive spores. *C. perfringens* is widespread and contamination of food during preparation is common. Type A2 tends to resist cooking during the preparation process due to its heat resistance. Type A1, which is less resistant, tends to be introduced by faulty catering between cooking and serving. The organism is found widely in raw foods and, although cooking destroys the vegetative cells, some spores may survive and typically the organism would be found in foods such as cooked meats which had been refrigerated inadequately. Due to the high incidence rate of *C. perfringens* in foods, qualitative results (positive or negative) need to be substantiated with quantitative results. Normally counts would not exceed $10^{4-5}$ $g^{-1}$, but in cases of food poisoning counts of $10^{9}$ $g^{-1}$ may be found. It has been suggested that food poisoning caused by *C. perfringens* requires the ingestion of $10^{7-8}$ viable cells. Quantitative counts may be carried out on blood agar containing the antibiotic neomycin.

*C. botulinum* (Figure 7.5) is the cause of a rare form of food poisoning caused by eating food in which the organism has multiplied and formed toxin. The protein toxin is extremely dangerous and death frequently results due to paralysis of the respiratory system. A number of different types of toxin are found. There have been a number of outbreaks this century, but the condition is rare in the UK, although a recent case involved hazelnut yoghurt. An outbreak always receives considerable media attention due to the high mortality rate. The most common cause of this type of food poisoning is faulty processing during the preparation of the food, thus allowing the organism to multiply prior to ingestion. When the organism grows on a highly proteinaceous food, such as canned meat or fish, the tin is usually badly blown and the proteolytic nature of the organism usually produces an extremely foul smelling product. The organism is also capable of growing on low protein, low acid foods, and in these conditions the proteolytic nature is less obvious.

A number of media are available for the growth of clostridia. Cooked Meat medium is widely used and allows differentiation of clostridia into proteolytic (protein digesting) and saccharolytic (sugar splitting) types. It is suggested that Universal bottles containing medium should be nearly full to reduce the air space inside the bottle. Once inoculated, the tops should not be screwed on too tightly

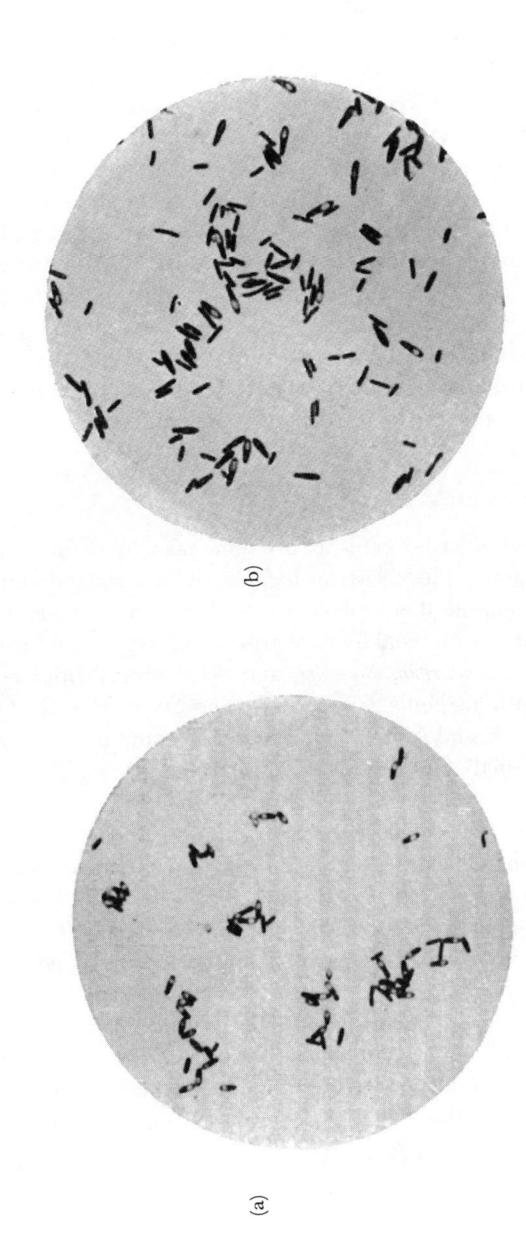


**Figure 7.5** (a) *Clostridium botulinum* (× *1000*). (b) *Clostridium butyricum* (× *1000*) *showing subterminal spores*
(Reproduced by permission from 'Topley and Wilson's Principles of Bacteriology and Immunity', 4th Edn., ed. G.S. Wilson and A.A. Miles, Edward Arnold, 1957)

as clostridia produce quantities of gas. Saccharolytic types produce acid from the glucose present, and the medium may appear pink/red. Proteolytic types produce considerable amounts of $H_2S$, which causes blackening of the medium.

There are a number of other media suitable for the growth of clostridia. The most important are Reinforced Clostridial Medium (RCM), Iron Sulfite medium, and Thioglycollate medium. The aim of all these media is to provide an environment of sufficient reducing power to allow anaerobes to grow in the presence of air. Anaerobes will grow on many media if cultured in an anaerobic jar, but this is not a simple process and should be avoided if possible.

RCM allows the growth of numerous anaerobes and is suitable for the counting of clostridia from a wide variety of foods. Iron Sulfite Agar allows the growth of a variety of anaerobes at elevated temperatures (55 °C). It is particularly suitable for the growth of thermophilic anaerobes producing sulfide spoilage. Thioglycollate medium is also used for the growth of many anaerobes. It is widely used for testing preparations containing mercurial compounds used as preservatives, as these are neutralised by thioglycollate. Thioglycollate should be freshly prepared before use, and the instructions of the manufacturer should be followed carefully.

### 3.3 Actinomycetes

This group of micro-organisms are not easy to define and several different systems are found. These systems include all bacteria containing a high guanine/cytosine content in their DNA, or all bacteria showing true branching and tending towards a mycelial form of growth, whilst a third includes genera such as *Actinomyces, Mycobacterium, Nocardia*, and *Streptomyces*. Whilst all the genera are of medical importance, only *Mycobacterium* and *Streptomyces* need to be considered in the context of this book. The only characters common to all of these organisms is that they are small, Gram-positive, non-motile rods.

*3.3.1 Mycobacterium.* This genus are Gram-positive, non-motile rods that do not show branching. They are acid-fast, do not form endospores and do not produce aerial hyphae and spores. They attack sugars by oxidation, but may take a long time to do so as many of them are very slow growing.

The most important organism in this genus is *M. tuberculosis*, the causative organism of tuberculosis. In the past, this organism was commonly found in raw milk and dairy products produced from infected cattle. This is now rare in the UK due to the regular screening of dairy herds. It should be noted that the incidence rates of TB in developed countries have started to rise recently after many years of decline. This is due to a number of factors, and food handling companies should always be aware of the possibility of carriers amongst their staff.

*M. tuberculosis* can be grown on Lowenstein–Jensen medium which contains fresh eggs and glycerol. The eggs supply the fastidious growth requirements of the organism and glycerol suppresses the growth of many other strains of *Mycobacterium*. Cultures are incubated at 35 °C and should be examined regularly for periods of up to eight weeks. Plates incubated for this length of time will dry out, so it is a useful idea to seal the edges of the plates with cellotape. Pyruvic acid egg medium may also be used, the pyruvic acid enhancing the growth of *Mycobacterium*. Due to the slow growing nature of *M. tuberculosis* and the difficulty in differentiating it from other non-medically important *Mycobacterium* species, it is suggested that any samples suspected of containing this species should be referred to a specialist laboratory.

*3.3.2 Streptomyces.* These are an important group of organisms industrially, due to their ability to produce a variety of antibiotics. They can also cause spoilage by producing musty or earthy smells, which can be absorbed by foods stored nearby. They form a Gram-positive mycelium producing aerial hyphae carrying short chains of spores, and are strongly aerobic. Some species are pigmented.

When grown, the morphological appearance of this genus may cause problems as the mycelium tends to be buried in the agar and cannot be removed easily for staining. Thus, when a loopful of organisms is taken for staining, a greyish mycelium is left on the agar and only the spores are removed. These appear as Gram-positive cocci. Any culture behaving in this manner should be suspected of being *Streptomyces*. An additional quick test for Streptomyces is to smell the plate carefully; a definite smell of soil is strongly indicative of *Streptomyces*.

*Streptomyces* will grow on a wide variety of media but is frequently outgrown by numerous other genera. They may be grown more selectively on media used for the detection and counting of fungi. One such medium is Czapek Dox agar which uses nitrate as sole nitrogen source, thus eliminating the more demanding bacterial species.

## 4 GRAM-NEGATIVE BACTERIA

### 4.1 Gram-negative Cocci

There are a number of groups of Gram-negative cocci but none of these are of significance in this book.

### 4.2 Gram-negative Rods

There are a very large number of Gram-negative rods (Figure 7.6 and also Table 2 on p. 75) of importance in industrial and food microbiology. Many of these have very similar properties and differentiating between them can cause major difficulties.

*4.2.1 Brucella.* These organisms are pathogens capable of causing low level fever symptoms (undulant fever), whose cause is very difficult to pinpoint. They have also been implicated in abortions in certain species. They may be found in milk from cattle, goats, and sheep, and also in unpasteurised cheese made from such milk. They are short, Gram-negative rods, which are non-motile and aerobic, although some strains grow better at increased levels of carbon dioxide.

*Brucella* are very fastidious organisms requiring a rich growth medium. They are also slow growing, and media are therefore supplemented by a cocktail of

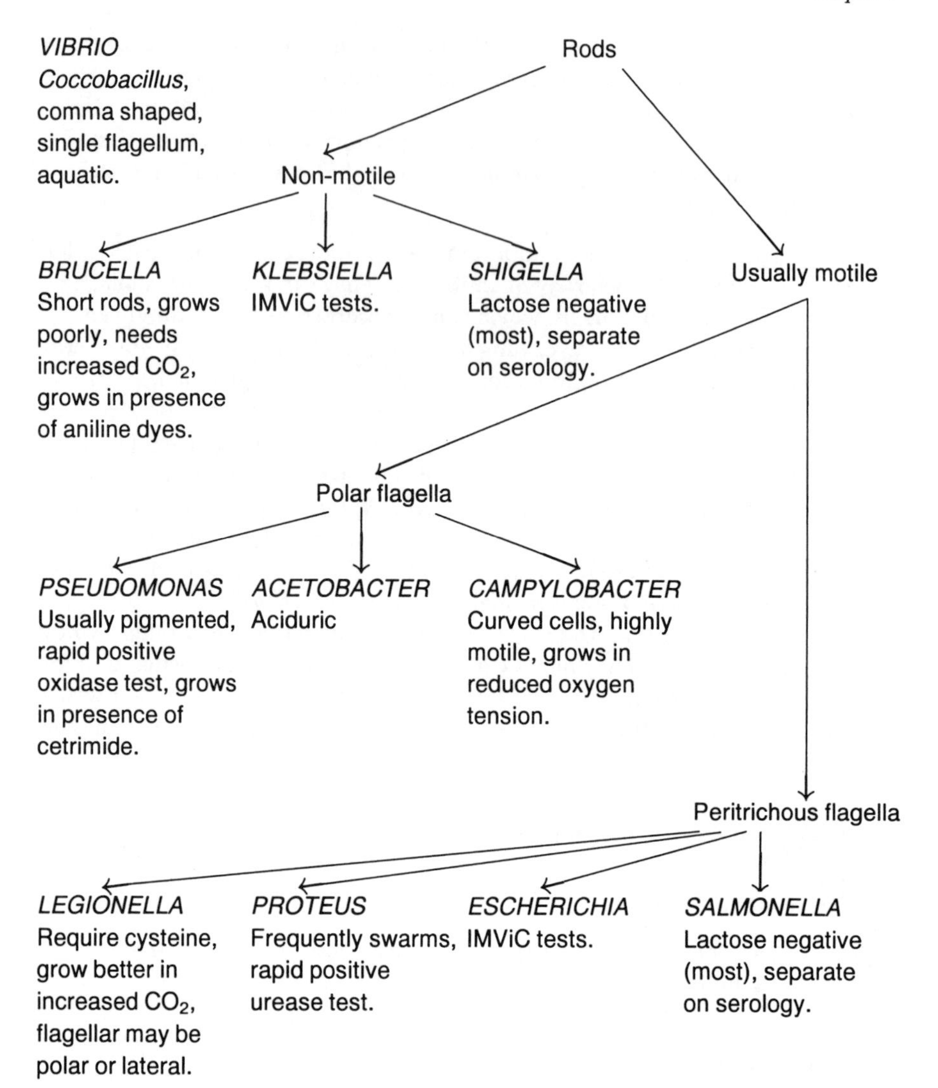


**Figure 7.6** *Gram-negative rods and coccobacilli*

drugs designed to suppress the growth of other species. Some selective media also include bacteriostatic dyes such as gentian violet, but this has been shown to be inhibitory to certain strains of *Brucella*. Suitable selective media are sold by a number of companies.

*4.2.2 Acetobacter.* The organism *Acetobacter aceti* is used in industry to convert ethanol to acetic acid in the production of vinegar. It is also found in spoilage situations in wine and beer, a process known as acetification, and also on damaged fruit. Spoilage may also be caused by *A. suboxydans,* a species forming copious amounts of slime. An enzyme from this genus is also used for converting the sugar alcohol sorbitol to sorbose in the industrial synthesis of Vitamin C.

They are short Gram-negative rods which may be motile, and which are always strongly aerobic and are capable of growth at pH values of 4.0. *Acetobacter* grows well on media supplemented with yeast extract, and containing an easily oxidisable carbon source such as ethanol which is converted to acetic acid, or glucose which is converted to gluconic acid. The medium may be rendered fairly selective by reducing the pH, which suppresses the growth of many competitors.

*4.2.3 Pseudomonas.* This genus is a very complex one which is found widely in spoilage situations, and also in a number of pathogenic situations from a wide range of host species. The group are referred to as opportunistic pathogens, in that they generally exploit a situation created by a primary pathogen.

Their ability to cause a wide range of spoilage is due to a number of features. These include their ability to use a large number of carbon and simple nitrogen sources, the production of off flavours, their proteolytic and lipolytic activities, slime production, pigmentation of some species, and the ability to grow at low temperatures.

They are Gram-negative rods, motile by polar flagella, and the majority give a strong oxidative reaction in O/F medium, although a small number may give an alkaline reaction. Some species are pigmented. They grow well at 37 °C and a number of the more important spoilage species will also grow at 5 °C, although few grow at 43 °C. They generally give a very strong oxidase reaction within a few seconds, which is widely used as a confirmatory test.

*Pseudomonas* grow well on a wide variety of media, frequently outgrowing other genera which may be present. They may be grown selectively by incorporating cetrimide and antibiotics such as nalidixic acid into the medium. The growth of the most important species, *P. aeruginosa*, is also considered in the chapter on water. A wide range of commercial media are available, many of which incorporate cetrimide.

*4.2.4 Achromobacter.* This genus is very poorly defined, and for many years was a dumping ground for all non-pigmented, Gram-negative rods which could not be identified. They have been implicated in 'off flavours' of eggs. Cowan and Steel[1] define them as aerobic Gram-negative rods, motile by peritrichous flagella.

*4.2.5 Chromobacterium.* This is also a poorly defined genus consisting of motile Gram-negative rods, which are either aerobic or facultative anaerobes, and which are pigmented. They have also been implicated in spoilage situations, and again tend to be a dumping ground for all pigmented Gram-negative rods, which cannot be positively identified.

*4.2.6 Vibrio.* This genus contains a number of medically important species found in contaminated water and foods. The species *Vibrio cholerae* is water-borne and is the causative organism of cholera (Figure 7.7), whilst *V. parahaemolyticus* is frequently associated with food poisoning from shellfish (Chapter 8). *V. cholerae,* causing epidemic cholera can be divided into two biotypes, the classical and El Tor, the El Tor type being the more commonly found in recent years and generally causing a lower mortality rate than the classical variety.

The organism is a motile, short Gram-negative rod, with a single polar

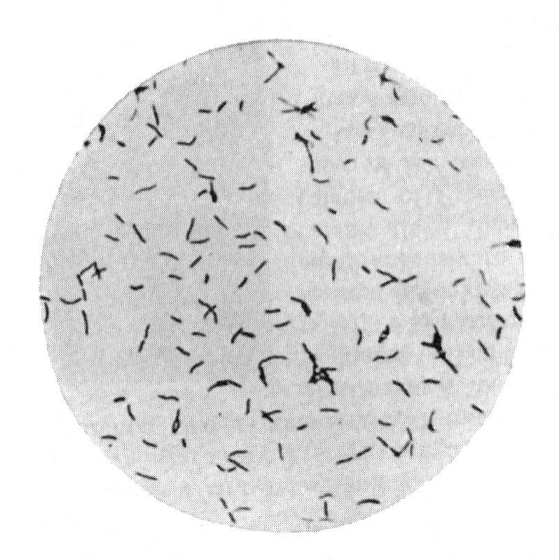

**Figure 7.7** *Vibrio cholerae (× 1000)*
(Reproduced by permission from 'Topley and Wilson's Principles of Bacteriology and Immunity', 4th Edn., ed. G.S. Wilson and A.A. Miles, Edward Arnold, 1957)

flagellum and, if stained, may appear 'comma' shaped. *Vibrio* spp. may be enriched by growth in alkaline media (pH 9.0) containing NaCl. Marine species such as *V. parahaemolyticus*, will grow at 6% NaCl, whilst all species will grow at 1% NaCl. A number of different serotypes of *V. cholerae* are found and identification of these should be carried out in a pathology laboratory. Deliberate isolation of *V. cholerae* should not be attempted in a routine analytical laboratory.

*Vibrio* species are difficult to isolate from natural sources as they tend to be present with coliforms, which outgrow them on many media. Several selective media are available, including DCLS Agar which contains sodium desoxycholate, lactose, sucrose, and the indicator Neutral Red, the desoxycholate suppressing the growth of coliforms. Cholera medium works in a similar manner, in that high levels of ox bile are used to suppress the growth of the normal gut flora.

*4.2.7 Legionella.* The habitat and growth media for these organisms are considered in the chapter on water. They are Gram-negative rods, which require cysteine and iron salts for growth and, although aerobic, prefer an enriched carbon dioxide atmosphere.

*4.2.8 Alcaligenes.* These are found commonly as spoilage organisms in dairy foods, producing ropiness and slime growth. They are motile Gram-negative rods, capable of growth down to 10 °C and are strict aerobes which are able to use very simple nitrogen sources.

*4.2.9 Aeromonas.* This genus is found commonly in dairy foods and fish where it contributes to spoilage problems by the production of trimethylamine. They are Gram-negative rods, which may or may not be motile. The non-motile species, which will not grow at 37 °C, are pathogenic for fish, whilst the motile species, which grow at 37–41 °C, are found widely in water and have also been isolated from wounds and cases of diarrhoea.

*4.2.10 Campylobacter.* These are long, thin Gram-negative rods which frequently appear as spirals or short chains forming bird wings. They are highly motile and if studied by the hanging drop method may appear to move in a corkscrew manner. One species, *C. jejuni*, has been shown to be a major cause of food poisoning over the last few years. They grow best at reduced oxygen levels (5–10%), and do not grow well in standard media unless it is supplemented by blood or a similar additive. Media must also be supplemented with antibiotics to avoid *Campylobacter* being overgrown by the normal enteric bacteria. In the event of very low numbers of *Campylobacter* being expected, it may be advisable to add the sample to an enrichment medium, before subculturing it into the selective medium.

## 4.3 Enterobacteria

This group, as the name suggests, includes those genera found in the intestines. A number of genera have been excluded as they have not been specifically implicated in food poisoning or spoilage situations, although one or two of the 'laboratory nuisances' have been included, as they can cause confusion with more important genera. A very wide range of media are available for the growth of the Enterobacteria. Some of these are general media, but many have been formulated to allow the differentiation of the various groups, and will be dealt with on a genus by genus basis.

*4.3.1 Enterobacter.* This genus are motile Gram-negative rods, which attack sugars fermentatively to form gas. They may be found causing spoilage in milk.

*4.3.2 Erwinia.* These organisms are Gram-negative motile rods which are generally considered to be plant pathogens and may be found in spoiled vegetables and fruits. Some strains have also been found in animals and many of these are pathogenic for plants. The classification of the genus is generally very confused.

*4.3.3 Escherichia.* This genus is one of the most important groups and is considered in some detail in the chapter on water microbiology. A number of serotypes are associated with diarrhoeal infections in humans (enteropathogenic *E. coli*), and have been isolated from a variety of foods, usually those that have been inadequately cooked and/or stored. They are Gram-negative rods which may be motile and are either aerobic or facultative anaerobes, attacking sugars in a fermentative manner usually producing gas. A range of characters is found with some of the less typical varieties being difficult to distinguish from *Shigella*. The working classification, used in water microbiology for thermotolerant coliforms, is the production of acid and gas from lactose at 44 °C, and indole from tryptophan also at 44 °C.

*Escherichia* may be grown on MacConkey medium (either broth or agar), which is used in water testing. Eosin Methylene Blue (EMB) Agar is widely used and *E. coli* forms small discrete colonies with a greenish metallic sheen and a dark purple centre. *Enterobacter aerogenes*, a species with which *E. coli* is frequently confused, forms larger mucoid colonies which tend to coalesce. The metallic sheen is usually absent, and the centre is brown. Non-lactose fermenters produce colourless colonies on this medium. The medium also allows the growth of the yeast *Candida albicans*. Further tests should be carried out to confirm any presumptive identifications obtained on EMB agar.

Desoxycholate Agar is recommended by APHA for the counting of coliforms in dairy products. APHA[5] recommend placing a sample (1 ml) in the petri dish and preparing a pour plate of medium. Once this has set, a further 5 ml of uninoculated medium is poured over the plate. Plates are incubated at 35 °C for 18–24 hours and coliform numbers can be estimated by counting all dark red colonies. Enteric bacteria which are non-lactose fermenters form colourless colonies, whilst non-enteric bacteria are inhibited by the desoxycholate.

Several of the media mentioned above may be supplemented with methylumbelliferyl glucuronide (MUG). Hydrolysis of this releases methylumbelliferone which fluoresces strongly under UV light and gives increased sensitivity of detection of *E. coli*.

*4.3.4 Klebsiella.* This group of Gram-negative rods are non-motile and may be separated from *Escherichia* by means of the IMViC tests (Chapter 6).

*4.3.5 Proteus.* Organisms from this group are involved in the spoilage of eggs, meat, fish, and shellfish. They occur in large numbers in unrefrigerated foods, and it has been suggested that they may be implicated in food poisoning. They are Gram-negative rods, which are highly motile, having a tendency to swarm across agar plates and overgrow other organisms, making their identification difficult. Methods of dealing with this problem include increasing the level of agar in the plates and drying plates thoroughly. One useful test for the identification of *Proteus* is the urease test (Chapter 6), which frequently produces results within two hours. There is considerable controversy over the taxonomy of this group and organisms migrate into and out of the genus with considerable frequency.

*4.3.6 Salmonella.* This genus is one of the most important in food poisoning and has been implicated in many outbreaks of gastro-enteritis. They are Gram-negative rods which are usually motile, although there are important exceptions. There are a number of organisms in this group producing atypical reactions and a specialist text should be consulted if necessary. They may be subdivided into groups by their serology and by phage typing, although these techniques are beyond the capabilities of most analytical laboratories. Material implicated in an outbreak of food poisoning, which is thought to be caused by *Salmonella*, should be sent to a Public Health Laboratory.

As the genera *Salmonella* and *Shigella* are so important in food microbiology, there are a very large number of specialised media for the differentiation of these two genera, both from each other and other genera. Media containing sodium selenite is considerably more toxic to *E. coli* than to either *Salmonella* or *Shigella* species, and selenite broth is widely used for enriching *Salmonella/Shigella* from samples. Media containing selenite is recommended by the AOAC[6] for detection of *Salmonella* in eggs. Selenite is toxic and the manufacturer's recommendations for handling should be followed.

Media containing tetrathionate, which is also toxic to *E. coli*, is used in a similar manner to isolate *Salmonella* species. This medium does allow the growth of *Proteus*, which may be inhibited by the addition of the antibiotic novobiocin, thus rendering the medium specific for *Salmonella*. Tetrathionite broth may be used for the enrichment of contaminated material containing *Salmonella*.

Brilliant Green Agar may also be used to isolate and count *Salmonella* in eggs and dairy produce generally (APHA and AOAC). The growth of lactose fermenters such as *E. coli* is generally inhibited, but where they do grow, they form yellow/green colonies. Non-lactose fermenters (*Salmonella* spp.) produce pink/red colonies surrounded by an intense red zone. *S. typhi* does not usually grow on Brilliant Green Agar. *Proteus* and *Pseudomonas* may grow, and modified versions of this medium are available which suppress the growth of these genera.

Desoxycholate Citrate Agar may also be used as a selective medium for *Salmonella* and *Shigella*. Non-lactose fermenters form colourless colonies surrounded by an orange/yellow halo, whilst lactose fermenters form pink colonies, with the acid produced causing a halo of precipitated desoxycholate. The growth of *E. coli* is usually suppressed. Organisms forming a black centre to the colony are $H_2S$ producers, the black colour being due to the formation of iron sulfide. Although Desoxycholate Citrate usually suppresses the growth of *E. coli*, it does not kill it, and if colonies are subcultured into other media, *E. coli* may re-emerge.

Bismuth Sulfite Agar is also highly selective for the isolation and recognition of *Salmonella*, especially lactose fermenting *Salmonella*. The combination of Bismuth Sulfite and Brilliant Green in the medium usually suppresses coliforms, but allows the growth of *S. typhi* and other *Salmonella* spp. The sulfur containing compounds are reduced to sulfide which reacts with iron salts forming a black precipitate. After 24–48 hours incubation, *Salmonella* spp. are found as black colonies, frequently with a metallic sheen and surrounded by a black halo.

Salmonella/Shigella (SS) Agar is recommended by the AOAC for the isolation of *Salmonella* and *Shigella* from foodstuffs. Coliform and Gram-negative bacteria are inhibited by high levels of Brilliant Green and bile respectively. Non-lactose fermenters such as *Salmonella* and *Shigella* spp. produce colourless colonies, whilst lactose fermenters (resistant *E. coli*) produce red colonies. A number of species of both *Salmonella* and *Shigella* are late lactose fermenters, and growth on this medium should be continued for several weeks to check this feature.

Xylose lysine desoxycholate agar is also useful for distinguishing *Salmonella* and *Shigella*, *Salmonella* producing red colonies with black centres, and *Shigella* producing red colonies without black centres. Non-pathogenic coliforms produce yellow colonies.

Simmons Citrate Agar, which utilises citric acid as the sole carbon source, may be used to differentiate certain *Salmonella* groups from each other. It does not allow the growth of coliforms or *Shigella*. Bacteria able to grow on citrate produce an alkaline reaction changing the medium from green to bright blue. It is important to avoid the carry-over of medium with the inoculum, and the inoculum should either be diluted with sterile saline prior to inoculation, or alternatively, any presumptive positives should be inoculated onto a fresh plate.

Some *Salmonella* species do not conform to the general description of the genus.

A number of them only react to various sugars very slowly, especially lactose. Any apparently negative result obtained in sugar fermentations should be subjected to a prolonged incubation period before being scored negative. In addition, one of the most important organisms, *S. typhi*, produces a number of atypical results as well as producing an infection (typhoid), which is atypical of the genus. *S. typhi* may be distinguished from other *Salmonella* by the inability to grow on citrate as sole carbon source (Chapter 6) and the non-production of gas when grown on glucose. If there is any suspicion that a sample contains *S. typhi*, it should be referred immediately to a Public Health Laboratory.

The organisms occur commonly in various foods, but some of the most important sources of human gastro-enteritis are poultry, eggs, and products derived from these, especially if they are raw (meringues) or have been held in an unrefrigerated state for a long period of time. The number of organisms required to cause an infection ranges from about $10^{6-9}$ depending on the virulence of the causative organism and the resistance of the host. The onset of symptoms ranges from about 18–36 hours after ingestion of the infected food, unlike staphylococcal food poisoning, where the symptoms appear within 2–3 hours. Victims may remain in a carrier state for a considerable period of time (months) after the symptoms have vanished, and they are an important source of re-infection.

*4.3.7 Serratia.* Many organisms in this genus are pigmented, forming a bright red colour, and were responsible for the 'miraculous' appearance of spots of blood on holy bread. Some species are non-pigmented but most produce pigment on starchy foods. They are motile Gram-negative rods, which attack sugars in a fermentative manner, usually producing gas.

*4.3.8 Shigella.* These organisms cause severe dysentery and may be found in contaminated water, and also in a number of foods with a high water content, which have not been cooled sufficiently rapidly after cooking. They are Gram-negative, non-motile rods which can be sub-divided on the basis of the fermentation or non-fermentation of mannitol. Recent studies using DNA have shown that they are very closely related to *E. coli*. At the practical level they can be distinguished from *Escherichia* by their non-motility, their inability to decarboxylate lysine, their non-production of gas from glucose, and the inability of a number of strains to ferment lactose. Food samples suspected of containing *Shigella* should be referred immediately to a Public Health Laboratory.

## 5 FUNGI

Mycelial fungi are frequently found in food samples and spoilage yeasts are common in alcoholic drinks and fruit juices. Spoilage may be of several types, the most common being the formation of acids, mycotoxins, slimes and off flavours.

Detection of these species may be difficult as they tend to be present in small numbers initially, especially the spoilage yeasts in alcoholic drinks where only 2–3 organisms per litre may be present in the early stages. They also tend to grow more slowly than bacteria and are frequently overgrown.

A number of specialist media are available for the growth of fungi, but two common features are a low pH and the addition of antibiotics. Most bacteria grow best at a pH value of 7.0–8.0, whereas many fungi will grow at pH values of 5.0 whilst some yeasts will grow at pH 3.5. The antibiotics used are those such as chloramphenicol which inhibit the growth of procaryotic cells. In addition to these techniques, high levels of compounds such as sugars and glycerol, which reduce the $A_w$, may be incorporated. A low $A_w$ also tends to favour the growth of fungi and yeasts over bacteria.

Identification of these species is not easy[8] and a large scale or persistent problem is probably best referred to a specialist laboratory. An identification kit is available from API for the more common yeasts.

## 6 REFERENCES AND FURTHER READING

1. 'Cowan and Steel's Manual for the Identification of Medical Bacteria', 3rd Edn., ed. G.I. Barrow and R.K.A. Feltham, Cambridge University Press, 1993.
2. 'The Shorter Bergey's Manual of Determinative Bacteriology', 8th Edn., ed. J.G. Holt, The Williams and Wilkins Company, Baltimore, USA, 1984.
3. 'Oxoid, The Manual', 6th Edn., Compiled by E.Y. Bridson, Unipath Ltd., Wade Road, Basingstoke, UK 1990.
4. 'Difco Manual', 10th Edn., Difco Laboratories, Detroit, Michigan, USA, 1984.
5. 'Compendium of Methods for the Microbiological Examination of Foods', 2nd Edn., Compiled by the APHA Technical Committee on Microbiological Methods for Foods. ed. M.L. Speck, American Public Health Association, Washington, DC, USA. 1984.
6. 'Association of Official Analytical Chemists, Bacteriological Analytical Manual', 5th Edn., AOAC, Washington, DC, USA 1978.
7. 'Mackie & McCartney, Practical Medical Microbiology', 13th Edn., ed. J.G. Collee, J.P. Duguid, A.G. Fraser, and B.P. Marmion, Churchill Livingstone, Edinburgh, London and New York, 1989.
8. R.R. Davenport, 'Yeasts and Yeast-like Organisms', in 'Smith's Introduction to Industrial Mycology', 7th Edn., ed. A.H.S. Onions, D. Allsopp and H.O.W. Eggins, Edward Arnold, London, 1981.

CHAPTER 8

# Microbiology of Food

## 1 INTRODUCTION

The microbiological content of food can influence a number of factors including quality, public acceptability, shelf life (spoilage), and safety. The role of the analytical microbiologist is to determine whether the food meets certain requirements in terms of the number and types of micro-organisms present in the food. These requirements may be a standard, a specification, or a guideline. A standard usually has mandatory backing, whilst a specification is usually some commercially acceptable level, and a guideline is only advisory.

Historically, foods fall into two categories.[1] The first group is those foods with a long shelf life of weeks, months, or even years. These foods have been stabilised by techniques such as drying or smoking, or by the use of chemicals such as salt, sugar, or vinegar. The second group is perishable foods which are not shelf stable, and have a shelf life of only a few days or weeks. The first category is not usually subject to bacterial spoilage, and damage is generally caused by fungi (moulds) over a long period of time. The second category suffers from rapid bacterial spoilage and there is rarely any problem from slow growing fungi. The exception to this is fungal damage to fruit, fruit juices, and vegetables. These categories have become blurred over recent years. Sugar and salt are used less because of pressure from health groups, and health queries have also been raised about smoked foods.

There has been strong pressure from retailers to increase the shelf life of perishable foods. This has been achieved in several ways. There have been major advances in controlling contamination during the manufacturing and processing stages. The freezing and chilling of foods are also very widespread and have contributed greatly to the increases in shelf life. These methods are very effective, but when a fault occurs in the system it can lead to a very rapid increase in microbial numbers, especially of bacteria. In addition, the widespread sale of chilled foods has led to changes in the bacterial flora found, and it is now recognised that certain bacterial species which were once thought to be uncommon, can cause serious problems.

## 2 FOOD POISONING

Food poisoning may be caused by intrinsic agents such as allergens or toxic compounds present naturally, or by extrinsic agents which include micro-organisms. In the case of microbial food poisoning, the symptoms may be caused either by the growth of the micro-organism or by the production of some microbial toxic material. The organisms responsible may include bacteria, fungi, protozoa, or algae. An account of various types of food poisoning is found in Waites and Arbuthnott, who also cover legislation in both the UK and USA[2].

Food poisoning is a major disease globally, with infant deaths from gastro-enteritis world wide estimated in excess of $5 \times 10^6$ annually. Even in technically advanced countries, it is difficult to assess the true scale of food poisoning, as it has been estimated that there are 10–100 cases for every one that is actually reported. It is also difficult to quantify the problem as there are many variable parameters, *e.g.* what might be a few days discomfort for a healthy adult can be life threatening for an elderly or very young person. A further difficulty is that food poisoning micro-organisms may not produce the vomiting and diarrhoea that most people associate with this problem. Typical of this type of food poisoning are *Listeria*, associated with miscarriages, and *Brucella*, associated with prolonged fever symptoms.

## 3 MICRO-ORGANISMS IN FOOD

Four categories of micro-organisms may be present in food. The first is desirable micro-organisms which produce pleasing organoleptic changes in the food, *e.g.* yeasts in alcoholic drinks and certain bacteria and fungi in cheeses. The second group is spoilage organisms causing unpleasant organoleptic changes, *e.g.* sour milk. These are usually of little or no medical significance, but a few may cause intestinal problems if taken in large quantities. The third group is indicator organisms. These are species whose presence in food suggests conditions which might allow contamination by growth of or survival of pathogens. The assessment of indicators is especially useful where viruses may constitute a major health hazard. It is most unusual to isolate enteric viruses from food, and detection of viruses is generally carried out in a Public Health Laboratory after the infection. The fourth group is the pathogens causing gastro-intestinal problems. The medical symptoms may be due to a preformed toxin in the food being ingested, or from toxin produced in the intestinal tract, or from an infection, *i.e.* growth of a pathogen. The onset of symptoms is variable, ranging from 2–3 hours in the case of a preformed bacterial toxin, to days in the case of microbial infection. It can obviously be difficult to determine cause and effect when there is a period of days between the ingestion of the contaminated food and the onset of symptoms.

There may be some overlap between these groups. For example, *Staphylococcus aureus* may be just an indicator of generally unhygienic conditions, or it may produce staphylococcal enterotoxin causing very rapid symptoms (two hours) of gastro-enteritis. Similarly, *Escherichia coli* can cause an infection, be toxigenic, or just be an indicator of faecal pollution.

There may also be variations depending on the food. *Bacillus cereus* is a pathogen causing food poisoning when found in large numbers on rice, but is not generally associated with gastric problems when found in other foods, *e.g.* cream, when it is treated as an indicator organism.

Two types of microbiological limit may be recognised. The first is a limit of specific pathogens posing a health risk to the consumer. This is a relatively simple problem as the presence of dangerous organisms such as *Salmonella* spp., *Shigella* spp., and *Clostridium botulinum* is obviously not acceptable.

The routine testing for pathogens is usually confined to organisms presenting a moderate risk *e.g. Salmonella typhimurium*, *Bacillus cereus*, *Staphylococcus aureus*, *Campylobacter* spp., *Clostridium perfringens* and *Vibrio parahaemolyticus*. Routine testing for the following bacteria which are considered severe hazards should not be carried out in an analytical laboratory: *Bacillus anthracis*, *Salmonella typhi*, *Shigella* spp., *Vibrio cholerae*, *Brucella melitensis*, and *Clostridium botulinum*. If these organisms are found, the samples should be sealed immediately, all plates containing the sample should be destroyed by autoclaving, and the original material referred to a Public Health Laboratory immediately.

The second is a limit to indicator organisms which is a much more difficult problem to resolve. These indicator organisms include species such as *Bacillus cereus* and *Staphylococcus aureus* which occur more or less ubiquitously. In small numbers these organisms are not a problem, although they may proliferate under conditions of poor storage. In large numbers they suggest that the food was prepared under conditions of poor hygiene, and that spoilage or food poisoning problems will occur sooner rather than later. The counting of these organisms is therefore important: low numbers suggest a satisfactory situation, whilst high numbers suggest problems or potential problems. The problem of acceptable numbers is therefore a very subjective and emotional issue. This is considered later in this chapter and also in the chapter on counting.

As yet, there is no generally accepted agreement of standards of microbiological safety related to food. There are several definitions that are useful. Quality assurance ensures that the final product is of required quality for its intended purpose, and that this quality is consistent. Good manufacturing practice ensures that the process is fully defined before it starts, and that all necessary facilities are in place. This includes trained staff, suitable premises, and an adequate infrastructure covering storage, transport, and records. Quality control ensures that the necessary tests are made at each stage of production, and that the product is not released for public consumption until these tests have been passed.

## 4 TESTING

### 4.1 Hazard Analysis Critical Control Points (HACCP)[3]

The control of safety in the food industry is best achieved by the use of HACCP techniques. The technique was first used in the milk industry for pasteurisation to control tuberculosis, although it was not known by this name at the time. A critical control point is defined as a stage in the manufacturing process, which, if not controlled correctly, will cause a threat to safety or a spoilage problem. A hazard is defined as the potential to harm the consumer (safety) or danger to the product (spoilage). The overall responsibility for safety lies with senior management who should ensure that there are adequate resources available. The exact methods used will vary from company to company and product to product, but the principles are the same and are as follows. The basic approach to HACCP involves three factors. Firstly the determination of hazards, secondly the identification of the critical control points required to control the hazards, and thirdly the monitoring of the CCP.

The most important stage in approaching the problem is the preparation of an audit. This involves a list of ingredients and preparation of a flow diagram for each of the processes and products involved. The list of ingredients should include the raw material, any water involved in the process, and should also take account of materials obtained from alternative suppliers if the major supplier should fail. Other CCPs in the processing stage include the methods designed to eliminate or control the growth of micro-organisms, and staff handling of food.

The audit is also concerned with the chemical and physical characteristics of the product and conditions of storage, distribution, retailing and customer handling of the product. It must be appreciated that once the product has been sold, the producer has no further control over it and, when considering customer handling, it is advisable to consider a worst case scenario, and make due allowance for what might be termed the 'idiot' factor. The audit will include micro-organisms causing spoilage, in addition to those which are known safety hazards. Knowledge of the risk involved can be determined from food poisoning incidents known to have occurred with the same or similar products, and also from the chemical/physical characteristics of the products which might suggest hazards. The critical control points in the manufacture of cheese are shown in Figure 8.1.

Once the CCPs have been identified, the whole process can be reviewed with the objective of increasing the effective control, *e.g.* alterations in the time or temperature of cooking, speed of cooling, and temperature and time of storage (shelf life). There may be some conflict between safety and 'natural' products at this stage. A classic example is whether milk used for cheese should be pasteurised; a subjective improvement in the flavour if unpasteurised milk is used, compared with the destruction of pathogens and a reduction of spoilage organisms if pasteurised milk is used.

Further problems are the increased cost of additional safety measures related to the need to improve profit margins. Many users of HACCP systems define four levels of concern if there is a failure at a critical control point. These are as follows:

(a) High: a life threatening situation in the absence of positive control.
(b) Medium: a serious risk of food poisoning but not life threatening if positive control is absent.
(c) Low: a slight risk of food poisoning or spoilage if control is absent.
(d) None: no risk of food poisoning.

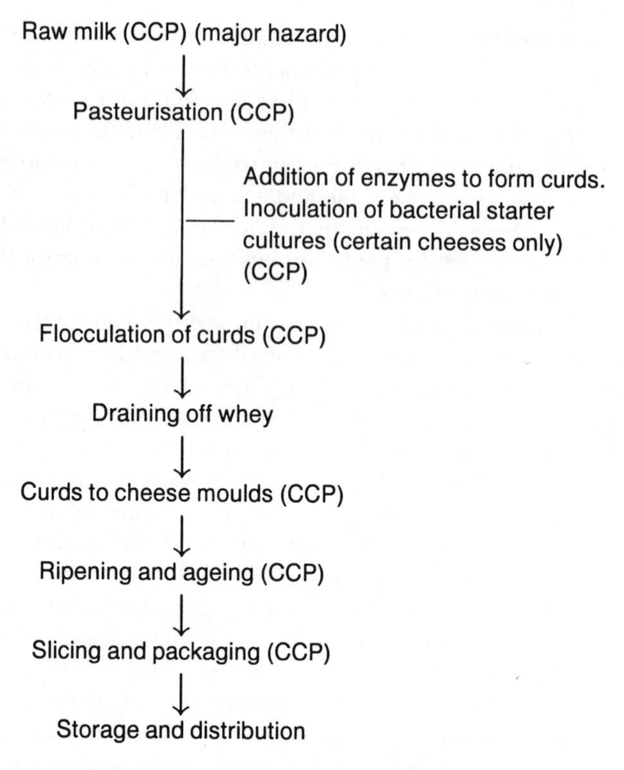


**Figure 8.1** *Critical control points in the manufacture of cheese*

Occasionally two differential levels of CCP are defined. Firstly, CCP1, which guarantees control of the hazard, and secondly CCP2, which reduces the risk but does not guarantee control. Monitoring the critical control points ensures that the process is proceeding correctly. The monitoring system should be sufficiently rapid to detect any malfunction before the product is released to the public. In the case of foods with a short shelf life, this is obviously a serious problem and is discussed in the chapter on counting. Monitoring should always be systematic with a sampling plan devised statistically.[4] The statistics of food sampling are also considered by Jarvis.[5] There needs to be adequate documentation of the whole process to ensure that, if a fault does arise, it can be located at the earliest opportunity, and the faulty batch withdrawn or recalled. This obviously implies that there is a method for responding rapidly to complaints.

## 5 MICROBIOLOGICAL ANALYSIS AND QUALITY CONTROL

Several major factors need to be considered. Firstly, have any micro-organisms present in the ingredients been destroyed? Secondly, if they have not been destroyed, are they capable of growth and/or toxin formation? Thirdly, is there any point in the process where micro-organisms could be introduced or reintroduced, and if such an introduction takes place, could they grow or form toxin?

Much of the laboratory time (perhaps 70–80%) will be taken up with routine samples taken from the processing line, and examination of these samples is designed to detect any problem. The preparation of samples for analysis takes a long time, therefore routine samples should be taken on days when analysis may be started and finished conveniently. Details of bottlenecks and problems are given.[6] The remainder of the time will be spent examining any samples involved in the outbreak of a foodborne disease and dealing with consumer complaints.

The basic requirements when sampling an unknown consignment (*e.g.* raw material of overseas origin) are the presence of pathogens and total count of indicator/spoilage organisms. In the case of pathogens, the requirement is that they should be absent, or not recoverable (which is not the same thing). It is obviously impossible to test the whole consignment, so the standard is that specific pathogens should not be recoverable from X grams of sample. In this case, the result is obviously pass/fail.

In the case of spoilage/indicator organisms, the requirement is that the food should not contain more than X bacteria per gram or ml. This obviously makes the assumptions that the consignment is homogeneous (or has been homogenised effectively) and that the sample is representative of the lot. When assessing spoilage/indicator organisms, there are three possibilities, pass, marginally acceptable or fail. The point at which these limits are set depends on the type of food, the processing to which it will be subjected, and the policy of the company purchasing/selling the merchandise. Decisions should not be taken on the basis of a single analysis. If a sample is microbiologically defective, then the analysis should be repeated several times and mean and SD values obtained. This is obviously essential if there may be litigation relating to the quality of the material.

There are several objectives for 'in house' quality control. Microbial testing of raw materials, intermediate, and finished products allows the effectiveness of microbial control during processing to be assessed, whilst another major consideration of quality control is the predicted shelf life of the product. Quality control requirements, *i.e.* what to test for, how frequently to test, and what is acceptable, are usually management decisions based on experience, and are not mandatory requirements applied by an outside body.[7]

The major tests used are the total aerobic viable plate counts at 22 and 37 °C on nutrient agar. Specialised media for yeast and mould counts may also be used.[8] High counts indicate contaminated raw material, contamination during the process or storage, or unsuitable processing conditions. High counts also suggest rapid spoilage will occur. Many foods have bacterial counts in excess of $10^6$ $g^{-1}$ at the time spoilage becomes evident. Anaerobic mesophilic (30 °C) plate counts may also be useful indicators to conditions which might favour the multiplication of anaerobic organisms such as *Clostridium* spp. Psychrophilic plate counts (5 °C) are useful as a prediction of the behaviour of food in cold storage.

Tests for coliforms or Enterobacteriaceae are useful for monitoring the effectiveness of mild heating (pasteurisation) and as an indicator of recontamination. The presence of coliforms does not indicate faecal pollution as there are a number of sources of these organisms. However, high counts do suggest conditions are ideal for the proliferation of *Salmonella* spp. or other enteric pathogens. Some Enterobacteriaceae must be checked in raw materials, especially if the processing will not destroy them. *Salmonella* spp. should be checked in all foods which are not processed by heat and which have a history of *Salmonella* contamination, *e.g.* mayonnaise, dried milk and desiccated coconut.

*Enterococcus (Streptococcus) faecalis* is not widely used as an indicator in food microbiology. It is useful, however, as it has a relatively high resistance to high temperatures, freezing, drying and disinfectants/detergents. It may therefore be preferred as an indicator in certain situations, as it is more resistant than the Enterobacteriaceae.

*Staphylococcus aureus* is a common contaminant of the skin, nose and mouth of workers. Large numbers indicate that personal hygiene and temperature processing are inadequate. *S. aureus* isolated from food is generally assumed to be toxigenic if it is coagulase positive. The problem foods for this organism are cooked foods which have been contaminated during processing. This takes place after the cooking stage, as the competition has been eliminated. Such contamination implies faulty handling. They are also able to grow at low $A_w$ and are osmophilic growing in products high in sugar and salt. Typical problem foods are dairy products (*e.g.* dried milk), cooked and cured meats, and artificial cream.

*Bacillus cereus* is found in rice meals which have been cooked and then not stored correctly at a low temperature. It is frequently associated with 'take away' meals which have been cooked and left at room temperature for some hours. *Bacillus* spp. are also of importance when examining defective canned foods.

## 6 SPECIFIC PROBLEM FOODS

It is obviously not possible to give a comprehensive list of foods in a single chapter, so only a number of examples will be considered. Further examples are found in references 4 and 9.

Bacteria present in food are either present on the surface or in the body of the food. If the bacteria are present in the body of the food, then 10 g of food should be weighed aseptically into a sterile blender and homogenised with 90 ml of sterile diluent, such as Ringers solution. This gives a one in ten dilution and further dilutions using sterile diluent may then be prepared. Details are given in the chapter on counting.

Foods such as intact vegetables, which are contaminated on the surface, should be weighed (100 g) into a sterile container and 100 ml of sterile diluent should be added. The sample should be shaken well for 20 minutes, and the assumption is made that the diluent has removed all of the surface bacteria. Dilutions can then be made as required.

### 6.1 Eggs and Poultry

The major organisms of concern in eggs are *Salmonella* spp. The level of contamination depends on several factors, such as age, porosity of the shell, the rate of cooling after laying, whether the egg is cracked, and the storage conditions. A major factor in the incidence of eggs in food poisoning is whether the eggs are cooked. Outbreaks of food poisoning due to the use of raw eggs are well documented.

The eggs should be broken and the contents homogenised aseptically. One volume of homogenate should be added to five volumes of sterile culture medium to dilute the enzyme lysozyme, which dissolves bacterial cell walls. The preparation is then tested for *Salmonella*.

Foodborne disease is frequently associated with the consumption of meat or poultry. The most important bacteria are *Salmonella* spp., which inhabit the intestinal tract of many animals, and contaminate meat after slaughter. A common cause of the problem is cross contamination of raw to cooked meat, which may be caused by the use of the same chopping block or utensils. Poultry is highly susceptible to infection by *Salmonella*, especially when reared in battery conditions, and poultry which has been incorrectly thawed or undercooked is a major cause of food poisoning incidents.

It is recommended[9] that chicken should be examined for *Salmonella* by placing it in a sealed plastic bag with 100 ml of sterile water. The chicken should be allowed to thaw (if frozen) and washed thoroughly with the water. The liquid should be collected, added to an equal volume of *Salmonella* double strength enrichment medium, and tests for *Salmonella* carried out.

Poultry may also be subjected to a surface examination as follows.[9] A swab is rubbed over 16 $cm^2$ of skin on the bird's breast and rinsed in 10 ml of sterile diluent. The resulting suspension is used for dilutions and total aerobic counts are carried out at 22 °C on nutrient agar, at 30 °C on MacConkey agar, and at 37 °C on blood agar for staphylococci and enterococci. The suggested results are as follows $<250\,000$ $16cm^{-2}$ for the total count, $<1000$ $16cm^{-2}$ for coliforms, $<5000$ $16cm^{-2}$ for enterococci, and $<100$ $16cm^{-2}$ for *Staphylococcus aureus*. A low total count indicates only a small number of spoilage bacteria are present, whilst low coliform and enterococci counts suggest slaughtering, hygiene, and processing techniques are good.

## 6.2 Meats

Vacuum packed meats are generally stored at low temperatures. Oxygen is absent and the meat is high in salt or nitrate. The outside of the packet should be swabbed with ethanol prior to opening with sterile scissors. An aliquot should be weighed out into a sterile container and macerated in sterile nutrient broth. Dilutions up to $10^{-3}$ should be prepared and plate counts should be carried out on blood agar for *Staphylococcus* spp. at 37 °C for 18–24 hours, and on MacConkey agar under the same conditions for coliforms. A total aerobic count of less than 1000 $g^{-1}$ should be obtained.[9]

## 6.3 Shellfish

Shellfish are also a common source of food poisoning frequently caused by

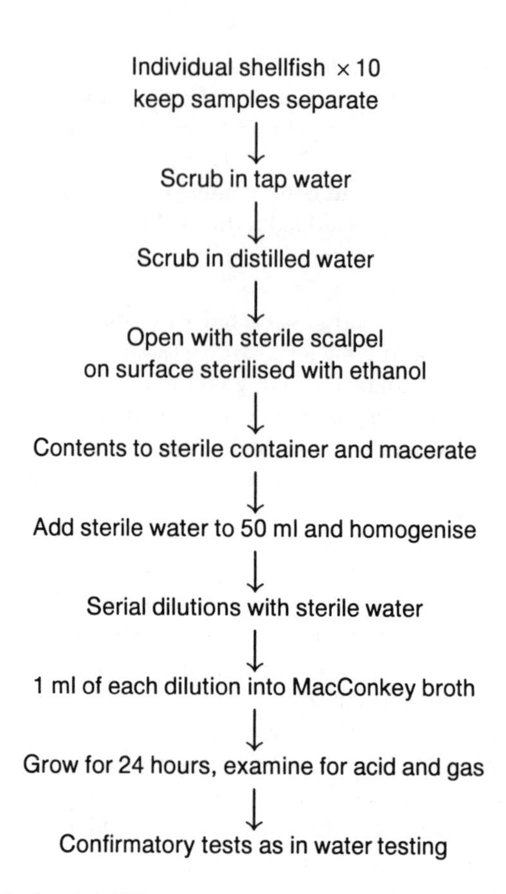


**Figure 8.2** *Examination of shellfish*

*Vibrio parahaemolyticus*. The suggested examination for shellfish is as follows[9] (Figure 8.2). Ten individual samples should be scrubbed in tap water then washed in sterile water. They should then be opened on a sterile surface aseptically and the contents placed in sterile containers keeping each shellfish separate. Macerate with sterile scissors, add sterile water to 50 ml and make ten fold dilutions. Inoculate 1 ml of each dilution into tubes of MacConkey broth containing a Durham tube and incubate for 24 hours at 37 °C. Confirmatory tests for coliforms should be carried out on all tubes showing acid/gas (for details see water testing). The shellfish may be considered satisfactory if coliforms are absent from at least 8 of the 10 tested. This test does not eliminate shellfish poisoning caused by dinoflagellate toxins.

Cooked shrimps/prawns have also been implicated in food poisoning. Aliquots should be weighed into sterile diluent and dilutions prepared which are used for aerobic plate counts.[3] Counts of $10^5\ g^{-1}$ at 22 °C are acceptable whilst counts of $10^6\ g^{-1}$ should be considered suspect. Samples should also be tested for coliforms.

## 6.4 Heat and Eat Meals

Frozen meat pies and plated meals usually give a relatively low total count at 22 °C, *i.e.* $< 100\,000\,g^{-1}$, but should be tested for coliforms and *S. aureus*. Frozen foods generally should be plated onto a rich medium to avoid the problem of cold stressed bacteria not growing. Spoilage psychrophiles should be incubated at 5 °C for up to seven days and other organisms should be counted at 22 or 30 °C for up to three days.

## 6.5 Vegetables

Frozen vegetables may have total counts of up to $100\,000\ g^{-1}$. Exotic imported vegetables should be tested for coliforms, *E. coli* and enterococci as human waste is widely used as a fertiliser in Asia. Cress has also been implicated in a number of food poisoning incidents and should be tested for *E. coli*, as it may have been grown in water polluted with sewage.

## 6.6 Tinned Foods

Spoilage in canned foods is due to micro-organisms surviving the heating process during canning or the introduction of organisms into the can after heating. Canning involves treading a fine line between killing the organisms responsible for food poisoning and spoilage, and not destroying the texture and flavour of the food.

A number of cans should be examined from the suspect batch. The outside of each should be examined for defects, especially in the seams, and any dented areas. The can should also be examined for blowing, that is whether the ends of the can are swollen due to pressure of gas produced by bacteria.

The outside of the can should be sterilized by flaming or washing with alcohol, and then opened with a sterile can opener. A sample of the contents should be removed carefully, especially if the can has blown. The samples should be diluted in sterile diluent as appropriate and incubated both aerobically and anaerobically for 2–7 days at the following temperatures, 5, 22, 37, and 55 °C.

The presence of members of the genera *Bacillus* or *Clostridium* suggest faulty processing as both these genera possess heat resistant spores. The presence of cocci and Gram-negative rods, which are not heat resistant, suggests faulty cans allowing the ingress of bacteria.

Dextrose tryptose bromocresol purple agar may be used for the growth of *Bacillus*, whilst a Reinforced Clostridial medium is used for *Clostridium*. If the can is damaged, then leakage is indicated and samples should be grown on Blood Agar and MacConkey Agar. If the sample smells of $H_2S$ then it should be subcultured onto Iron Sulfite Agar.

# 7 REFERENCES AND FURTHER READING

1. A.P. Williams and A. Bialkowska, 'Computer Assisted Identification of

Moulds' in 'Rapid Microbiological Methods for Food, Beverages and Pharmaceuticals', Society for Applied Bacteriology Symposium, Vol. 25, ed. C.J. Stannard, S.P. Petitt and F.A. Skinner, Blackwell, Oxford, 1989.
2. W.M. Waites and J.P. Arbuthnott, 'Foodborne Illness,' A Lancet Review, Edward Arnold, London, 1991.
3. A.H. Varnam and M.G. Evans, 'Foodborne Pathogens', Wolfe Publishing, Aylesbury, England, 1991.
4. 'Compendium of Methods for the Microbiological Examination of Foods,' ed. M.L. Speck, 2nd Edn., American Public Health Association, Washington, 1984.
5. B. Jarvis, 'Statistical Aspects of the Microbiological Analysis of Foods', *Progress in Industrial Microbiology* **21**, 1989.
6. M.K. Refai, 'Microbiological Analysis'. Food and Agriculture Organization of the United Nations, 1979.
7. 'Foodborne Micro-organisms of Public Health Significance', 4th Edn. ed. K.A. Buckle, J.A. Davey, M.J. Eyles, A.D. Hocking, K.G. Newton and E.J. Stuttard, Australian Institute of Food Science and Technology Ltd., 1989.
8. 'Food Mycology, A Guide to the Use of Oxoid Culture Media', Oxoid, 1991.
9. 'Mackie and McCartney, Practical Medical Microbiology', 13th Edn., ed. J.G. Collee, J.P. Duguid, A.G. Fraser and B.P. Marmion, Churchill Livingstone, 1989.
10. 'Identification Methods in Applied and Environmental Microbiology,' ed. R.G. Board, D. Jones and F.A. Skinner, Society for Applied Bacteriology, Vol. 29, Blackwell, 1992.
11. 'Isolation and Identification Methods for Food Poisoning Organisms', ed. J.E.L. Corry, D. Roberts and F.A. Skinner, Society for Applied Bacteriology, Vol. 17, Academic Press, London, New York, 1982.

CHAPTER 9

# Microbiological Analysis of Water

## 1 INTRODUCTION

The microbiological safety of drinking water is of great importance, as a number of important bacterial (*e.g.* typhoid, cholera, dysentery and Weil's disease) and viral (*e.g.* polio and hepatitis) diseases are water-borne. Many large cities and towns draw drinking water from rivers upstream and discharge it downstream. There may be several towns on a river and the last one may be using water that has been recycled by several communities above it. It is obvious that waste water containing sewage must be treated adequately before consumption to remove all pathogens and avoid outbreaks of the diseases mentioned above. In addition, the water must also have a low BOD (biological oxygen demand) and COD (chemical oxygen demand).

These aims are achieved by several methods. Initially there is a screening process to remove solid matter. This may be followed by an activated sludge process in which the liquid is actively aerated by stirring, or filtration through gravel and sand, or an oxidation pond where sewage is held for about a month before being released. Drinking water or water used in food manufacture is then further treated before consumption, usually by chlorination. A useful account of water microbiology is given by Olson and Nagy.[1] The microbiological examination of drinking water is covered in considerable detail in several books.[2,3] The HMSO report details the following:

(a) The principles upon which the examination of drinking water is based.
(b) The frequency with which drinking water should be examined bacteriologically.
(c) The type of bacteriological examination required.
(d) The interpretation of results.
(e) A guide to the Directives of the European Community.

The report details the monitoring of public supplies of water, and covers collection, treatment, and distribution.

Water supply companies have a legal requirement to supply 'wholesome' water and Regional Water Authorities have a statutory duty (various Public Health and Water Acts) to ensure that water is 'wholesome and sufficient'.

The Water Acts and the European Commission Directives[4] relating to the Quality of Water intended for Human Consumption give guidelines and suggest maximum values for a number of parameters, including microbiological ones. These directives also included, for the first time, private water supplies from boreholes, wells, and springs as well as public supplies. They also suggest a minimum sampling programme based on the volume of water supplied and the size of population served.

## 2 THEORY

The greatest threat to water supplies is contamination by raw sewage which contains pathogens capable of causing a number of diseases. The examination of drinking water for the presence of specific pathogens, whether they are viral, bacterial or protozoal, is time consuming, impractical, expensive, and requires a fairly high level of technical expertise. It is not necessary for routine testing and control.

Bacteriological examination of water for indicator organisms is the most sensitive method of testing for faecal pollution. Analysis for viruses requires sophisticated laboratory facilities and is unnecessary for routine analysis. Little information is available on the destruction of protozoa by water purification methods. The growth of pathogenic protozoa, such as *Giardia*, also requires sophisticated facilities, which would only be available in a specialised microbiology laboratory. The concern of the microbiologist is therefore not whether the water does contain pathogens but whether it could.

Pollution may occur at irregular intervals or at short notice, due to problems such as damage to distribution systems or heavy rain following a long drought. One satisfactory examination does not mean therefore that the water supply will remain safe over a long period of time. It is more reliable to examine water regularly using simple tests, rather than infrequently using more complicated ones. Over a period of time a microbiological 'feel' for a particular supply will be obtained and any deviation from this should be noticed immediately.

Monitoring for the presence of specific pathogens is unnecessary. Pathogens present tend to die out more quickly and are more difficult to detect than the normal human or animal gut microflora. It is also technically difficult to isolate and identify specific pathogens in the presence of the large numbers of other organisms found in contaminated water. Identification of pathogens requires techniques such as the serological testing of pure cultures. These are more suited to Public Health Laboratories and the isolation of certain pathogenic species, e.g. *Salmonella typhi*, is beyond the technical facilities of most analytical laboratories. The detection and identification of such species will not be considered in this chapter. Therefore the use of relatively rapid, simple, and cheap tests to determine the presence or absence of commensal indicator organisms is adequate. The bacterial indicators used for faecal pollution are *Escherichia coli*, the coliform group, faecal streptococci, and *Clostridium perfringens*. These indicator organisms persist much longer than pathogens and are relatively easy to isolate and identify as they are present in faeces in large numbers. Their presence in water shows that the water has been polluted by sewage at some time in its past history, and is therefore a potential health hazard. The tests for coliforms and *E. coli* are the most important of these routine tests.

## 2.1 Coliforms

The characteristic features of the coliforms are non-sporing, Gram-negative rods, which are bile tolerant and able to ferment lactose at 37 °C producing acid and gas within 48 hours. The term faecal coliform was used to describe coliforms capable of fermenting lactose to acid and gas within 24 hours at 44 °C. The term thermotolerant coliforms is now used more commonly in the UK, although the term faecal coliform is still widely used elsewhere. It should be pointed out that not all thermotolerant coliforms are faecal in origin.

The presence of coliforms in water indicates that pathogens could be present, and that the supply is potentially dangerous, although there is no correlation between numbers of coliforms and pathogens. The absence of coliforms indicates that pathogens are probably absent.

## 2.2 *Escherichia coli*

*E. coli* is a thermotolerant coliform (*i.e.* Gram-negative rod, non-sporing, bile tolerant, fermenting lactose or mannitol to acid and gas at 44 °C), which is also able to produce indole from the amino acid tryptophan; produces a positive Methyl Red test; is negative for the Voges–Proskauer test; and negative for the citrate test (see pages 59–62).

*E. coli* is very rarely found in water when faecal pollution is absent, and testing for *E. coli* and coliforms is the most sensitive method of demonstrating faecal pollution. Confirmation of the presence of *E. coli* indicates faecal pollution and the possible presence of intestinal pathogens. High counts suggest recent or heavy pollution, whilst low counts suggest slight pollution or pollution at some time in the distant past.

There is no quick and simple laboratory method to distinguish *E. coli* of human origin from *E. coli* of animal origin. However, this is not important as many animals, including rodents and birds, carry pathogens capable of causing human disease, and water carrying *E. coli* is suspect irrespective of the source.

The absence of *E. coli* combined with the presence of coliforms is more difficult to interpret. Whilst faecal pollution is the most probable explanation of contamination by coliforms, there are also other sources of coliforms which are generally innocuous, e.g. decaying vegetation and other organic matter such as washers and grease used in pipe joints.

## 2.3 Faecal Streptococci

This problem may be solved by examining the water for faecal streptococci. Faecal streptococci survive longer in water than *E. coli*, and are more resistant to chlorination. The presence of faecal streptococci is always indicative of faecal pollution and therefore they are a useful method of determining the significance of a result in which *E. coli* is absent, but coliforms are present.

Faecal streptococci are Gram-positive cocci usually growing in short chains and producing small colonies when grown aerobically. They grow in the presence of sodium azide and bile salts. Azide inhibits organisms containing a cytochrome $A_3$ system which is absent in streptococci.

Streptococci found in faeces belong to two major groups, one group found in humans, the other group occurring in various animals and not usually found in humans. The term faecal streptococci includes these species, all of which are in Lancefield's serological group D. *Streptococcus faecalis* is the species found mainly in humans. In certain countries, a comparision of *E. coli* numbers with the numbers of different streptococcal species may be used as an indicator of the source of the pollution. The faecal streptococci are more correctly classified now as the genus *Enterococcus*, the main species being *E. faecalis*. The term faecal streptococci is still widely used in many text books, especially on water microbiology.

### 2.4 *Clostridium perfringens*

This indicator organism is an anaerobic Gram-positive rod, usually present in faeces in low numbers. It produces spores which resist boiling, survive for long periods in water, and show considerable resistance to chlorination. The presence of *C. perfringens*, combined with the absence of coliforms and/or faecal streptococci, indicates that faecal pollution may have occurred at some time in the distant past. The presence of *C. perfringens* in water samples on several successive occasions suggests that the frequency of sampling should be increased. As *Clostridium* spores are resistant to chlorination, the presence of spores in contaminated water which has been treated, combined with an absence of coliforms, show that the treatment process has been successful.

### 2.5 Plate Counts

Aerobic plate counts may also be used as a general indicator of the condition of the water, although these counts have little significance in health terms. Plate counts are widely used in some countries. However, in the UK any changes in the colony counts are generally regarded as more significant than the actual aerobic plate count of any given sample. Changes in counts on plates grown at 37 °C, *i.e.* human body temperature, are regarded as more important than changes at 22 °C, which are liable to seasonal variation. A major use of plate counts is in assessing the effectiveness of water treatment processes, and in these conditions significant increases in bacterial numbers can give a warning of problems with the treatment process.

Large quantities of water are used in the food, drink, and pharmaceutical industries. This is frequently subjected to further treatment on entering the plant and is usually of very high quality. Plate counts are used widely in these industries to assess the effectiveness of this extra treatment. These industries also frequently test for the genus *Pseudomonas* especially the fluorescent pseudomonads, which are found widely in dust, air, and water, frequently occurring in spoilage situations. Their ubiquity means that they cannot be used as indicator organisms for faecal pollution. The pseudomonads are a diverse, poorly defined group and in this context tests for *Pseudomonas aeruginosa* are usually carried out. *P. aeruginosa* is a Gram-negative rod, aerobic, and giving a positive oxidase test (Chapter 6). Pigment production in the pseudomonads, although frequently quoted, is a variable characteristic and cannot be relied upon. A confirmatory test used widely for pseudomonads is their ability to grow on media containing cetrimide as a selective agent.

## 3 TESTING

The tests used to determine the presence of indicator organisms fall into two groups, multiple tube methods and membrane filtration methods.

### 3.1 Multiple Tube Methods

Multiple tube methods consist of adding known volumes of the water sample to tubes containing a specified volume of a selective medium in a liquid form (see Figure 9.1). The condition of the water sample may require sterile dilutions of the sample before the tests can be carried out. Once prepared, the tubes are incubated at the appropriate temperature for a period of time and tubes showing growth with an acid/gas reaction are scored as presumptive positives. The number of presumptive positives is checked against probability tables to give a presumptive coliform count.[2] The statistical calculations used in the probability tables (also known as McCrady's tables, McCrady, 1915) for calculating organisms in the multiple tube methods are based on the assumption that the micro-organisms found in the sample are evenly distributed and not clumped together.

Samples from the presumptive positive tubes are then inoculated into further medium for confirmation tests for coliforms and also to determine whether *E. coli* is present. Results from the Voges Proskauer, Methyl Red, and Citrate tests (see later) are used to confirm the presence of coliforms. Positive results from growth on MacConkey broth at 44 °C are scored to give a statistical probability of the Most Probable Number (MPN) of *E. coli* per 100 ml of water sampled.

Multiple tube methods are very sensitive to small numbers of micro-organisms due to the powerful amplification of these techniques. However, the methods are not very accurate statistically, and have considerable errors associated with them. In addition, the multiple dilutions needed are very demanding on technician time, sterile glassware and media. Tube methods may be run in conjunction with plate counts.

### 3.2 Membrane Filtration Methods

This method has been one of the major advances in water microbiology and is outlined in Figure 9.2. A measured volume of water is filtered through a cellulose

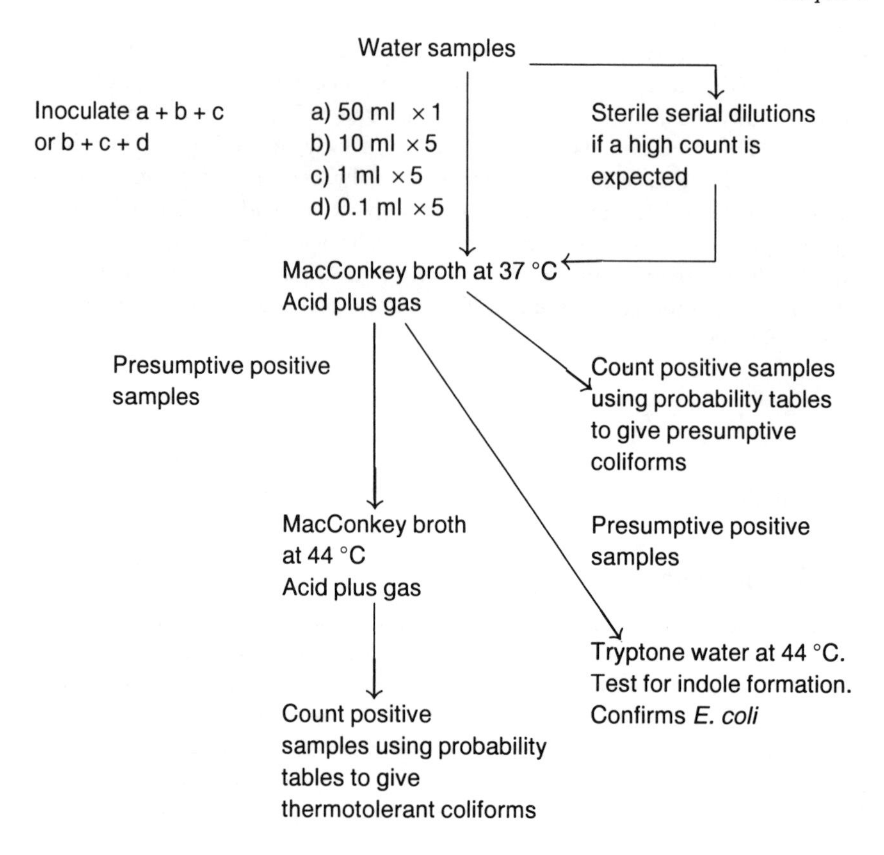


**Figure 9.1** *Water testing procedure for multiple tube method*

based membrane which has a pore size of 0.45 μm, enabling it to retain bacteria on the surface. After filtration the membrane is placed on a solid selective medium or on a cellulose pad saturated with a growth medium. The membranes are then incubated for an appropriate period of time and the number of colonies on the membrane are counted, frequently with the aid of a microscope, and results are expressed as number of micro-organisms per 100 ml of water. The surface of the membrane has a grid to facilitate counting.

The apparatus consists of a porous disc on a base attached to a vacuum source. Above this is a sterile funnel capable of holding varying quantities of water/sample. The funnel is attached to the base by clamps or magnets. The system is operated by removing the funnel, placing a sterile membrane filter on the base, replacing the funnel, adding the water sample to the funnel, and applying a vacuum. Water is sucked through the filter leaving the bacteria on the membrane.

The sample volume chosen should be such that between 10 and 100 micro-organisms are deposited on the filter. In the event that a volume of less than 10 ml is used, sterile diluent should be added and the sample thoroughly mixed before it is applied to the filter.

The membranes can be purchased ready sterilised or can be purchased non-sterile and autoclaved before use. The same is true of the absorbent pads. Sartorious sell the sterile pads impregnated with dehydrated medium in a sterile petri dish to which it is only necessary to add 3 ml of sterile distilled water before adding the membrane.

Membrane filtration counting has several advantages over multiple tube methods. It is ideal for handling large volumes containing only a small number of organisms of a certain type. As such, it is widely used in the wine and beer industries, where only small numbers of spoilage yeasts may be present but the produce will be stored for a long period of time. Samples for membrane filtration may be prepared more quickly than those for tube counting methods and do not need a large number of sterile pipettes and serial dilutions. However, it cannot be used to handle water or other samples containing large quantities of suspended material as this blocks the filters. In addition, the results may be more difficult to interpret, as membrane filtration methods do not demonstrate the ability of organisms to produce gas.

Both multiple tube and membrane filtration methods are not only qualitative; they are also quantitative, thus enabling an assessment to be made of the potential danger to health.

## 4 SAMPLING

The frequency of sampling varies considerably depending on the source and quality of the supply, size of population, volume of water distributed, and

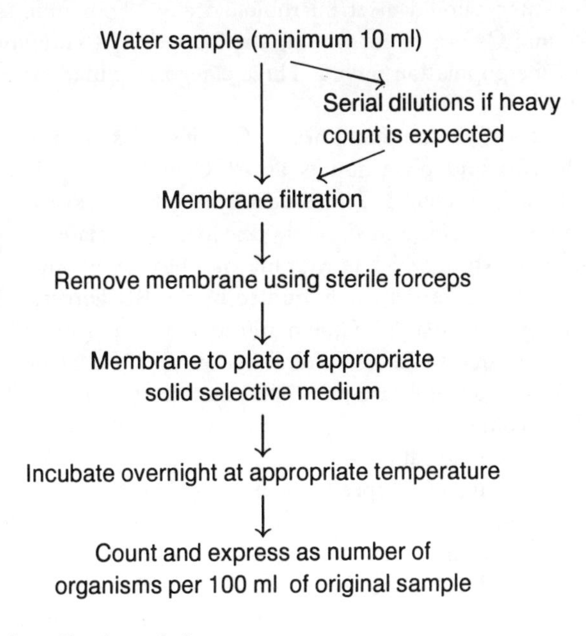


**Figure 9.2** *Membrane filtration method*

treatment processes used. An important factor is the health of the population. If there had been an outbreak of some water-borne disease, it would be necessary to increase the frequency of sampling significantly.

Sampling at the source of the water is generally of little importance. Sampling during the treatment process is not essential, but does indicate the effectiveness of the process control methods. The most important sampling relates to monitoring the distribution network, especially at the point of delivery, *i.e.* the tap.

The HMSO publication[2] suggests that ideally water should be monitored daily at each point where it enters the distribution network and that chlorine levels should be measured, preferably, on a continuous basis. The report recommends that where the supply serves a population of 10 000+, both of these requirements should be met. Where populations are below this level, the bacteriological quality may be monitored less frequently.

Water may become contaminated during distribution and this may vary seasonally (*e.g.* following drought or heavy rain). The water therefore needs to be examined regularly at various points in the distribution network. In the case of large populations (up to 300 000), it is recommended that one sample per 5000 population should be examined per month, with an increasing frequency of sampling as the population increases.

It is worth noting that where water is stored in holding tanks on site, especially if these are open, there may be considerable contamination from birds or rodents. Water that has been stored should be tested before use and it should not be assumed that because it was acceptable before it entered the holding tank, it is acceptable when it leaves.

The European Community Directive on water quality specifies guide values that surface waters should meet microbiologically.[4] Four monitoring categories C1, C2, C3, and C4 are given, relating the frequency of sampling to the volume of water and the population served. These categories differ somewhat to normal practice in the UK.

For example, minimum monitoring (C1) includes monitoring for thermotolerant coliforms and plate counts at 22 °C and 37 °C. These plate counts replace total coliform counts. This differs from UK practice which regards tests for presumptive coliforms and *E. coli* as essential, with plate counts at 22 °C and 37 °C as optional extras, which may or may not be carried out at the discretion of the Water Authority. Target values quoted by the EC Directive for plate counts are 10 organisms per ml at 37 °C, and 100 organisms per ml at 22 °C. Counts at 22 °C are carried out at 72 hours, and counts at 37 °C at 48 hours. This also differs from UK practice where the 37 °C count is at 24 hours. The UK attitude towards plate counts is that significant changes in plate counts are more important than absolute values.

Current monitoring (C2) specifies the number of times per year the water should be sampled for a given size of population. Periodic monitoring is current monitoring plus any further monitoring determined by the Authority, taking into account any factor influencing water quality. Occasional monitoring parameters are determined by the Authority with regard to the circumstances arising following an accident or special situations.

## 5 PRACTICAL SAMPLING

It is important that the sample should be representative of the water being examined. Certain precautions need to be taken to ensure that the water does not change significantly between the sample being collected and assayed.[2] These are:

(a) The sample bottle should be clean and sterile and should not be rinsed out before collecting the sample.
(b) The sample bottle should contain sodium thiosulfate to neutralise any chlorine present in the water. At the chlorine concentrations generally found in water supplies, the addition of 0.1 ml of 1.8% (w/v) sodium thiosulfate ($Na_2S_2O_3.5H_2O$) per 100 ml of water sample is adequate. Residual thiosulfate has been shown to have no significant effect on the *E. coli* or coliform counts.
(c) Care should be taken to avoid operator contamination of the sample whilst it is collected and subsequently analysed. The top of the sterile container should not be removed until the sample is ready to be collected and the container should be held at the bottom to avoid contamination by the hands. After the sample has been collected and the stopper replaced, the outside of the bottle should be dried.
(d) Water should be allowed to run through a tap for 2–3 minutes before sampling, to wash through any micro-organisms present inside the neck of the tap.
(e) The sample should be kept cool, and protected from light between being collected and examined.
(f) The sample should be examined as soon as possible after collection and certainly within six hours.

The HMSO report also makes recommendations for sampling from a variety of other sources such as reservoirs, wells and storage tanks.

## 6 TESTING

### 6.1 Volumes and Dilutions

The volume of water used to inoculate media depends on the expected levels of microbial contamination. In the case of counts on pour plates, 1 ml of water is placed in the plate and 9 ml of molten nutrient agar is added. The water may be used neat if it is of good quality, but should be diluted aseptically if of poor quality (see below). The agar should not be too hot (kills bacteria) or too cool as it will form lumps on pouring. The recommended temperature is 50–55 °C.

The volumes normally used for water of reasonable quality in tube counts are 50 ml of water inoculated into one 50 ml tube of double strength medium, 10 ml into each of five tubes containing 10 ml of double strength medium, and 1 ml into each of five tubes containing 10 ml of single strength medium. If the quality of the water is doubtful, then the 50 ml sample should be omitted, and be replaced with

0.1 ml samples inoculated into each of five tubes containing 10 ml of single strength medium. A fresh sterile pipette should be used for each inoculation.

If the water sample is seriously polluted, it may be necessary to dilute the sample before it is subjected to membrane filtration, tube inoculations, or plate counts. This is usually done as a series of serial dilutions. Serial dilutions are prepared using Ringers solution. A series of sterile tubes or bottles each containing 9 ml of sterile Ringers solution are prepared. One ml of sample is placed into the first tube in the series using a sterile pipette, and mixed thoroughly to give a dilution of $10^{-1}$. One ml of this is removed aseptically using a fresh sterile pipette, and added to the second bottle in the series to give a dilution of $10^{-2}$. This is again mixed thoroughly and the process is repeated until the required dilution is reached. The appropriate dilution is then used to inoculate the multiple tubes or is passed through the membrane filter. The objective of this process is to obtain dilutions which will give a mixture of positive and negative results when inoculated into multiple tubes, 10–100 colonies on membrane filters and 30–300 colonies when used for plate counts. The best method of establishing the appropriate dilution is by trial and error over a period of time.

## 6.2 Multiple Tube Tests

A number of media are suitable for testing for the presence of presumptive coliforms. These are widely available commercially. A commonly used one is MacConkey Purple Broth, which contains the disaccharide lactose, bile salts, and Bromocresol Purple (Figure 9.3). The principle of the test is that only organisms of the intestinal tract will grow in the presence of bile salts. Coliforms will grow and ferment lactose to acid and gas at 37 °C. The acid changes the colour of the indicator to yellow. Gas is collected in an inverted Durham tube which was initially filled with medium. A positive result is acid *and* gas (the quantity should be at least sufficient to fill the rounded end of the tube). The medium may be prepared in a single or double strength form.

There are numerous variations on this medium using mannitol instead of lactose, teepol or sodium lauryl sulfate instead of bile salts, and alternative indicators such as Phenol Red. However, all the tests are based on the ability of coliforms to grow in the presence of some inhibitor, and ferment lactose or mannitol to acid and gas, the acid changing the colour of the indicator.

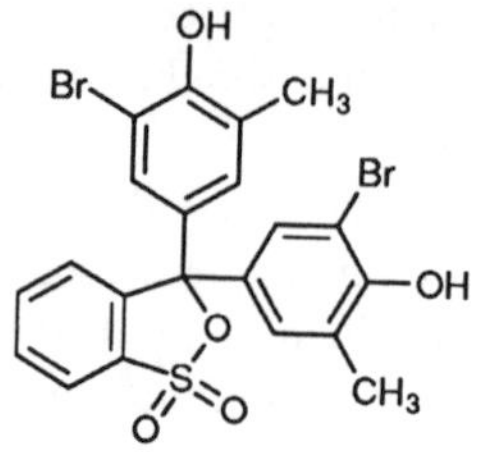

**Figure 9.3** *Structure of Bromocresol Purple*

The differential tests make use of the positive samples obtained in the presumptive tests. These differential tests are intended to differentiate *E. coli* from non-thermotolerant coliforms, and coliforms in general from non-coliforms. Samples from each of the positive presumptive tubes are inoculated into each of the four following media:

(a) A second tube of MacConkey Purple Broth which is incubated at 44 °C ± 0.5 °C. This may be replaced by alternative media, such as Brilliant Green Bile Lactose Broth. The temperature is critical and incubation should be carried out in a water bath;
(b) Tryptone Water at 37 °C;
(c) MRVP (Methyl Red Voges–Proskauer) medium at 37 °C;
(d) A slope of Simmons Citrate Agar at 37 °C.

The combination of (b), (c), and (d) is known under the acronym IMViC.

Tube (a) is examined for the production of acid *and* gas within 24 hours. A positive result is given by *E. coli* and a negative result by non-thermotolerant coliforms.

Tube (b) is tested for indole by the addition of Kovacs reagent (page 62). *E. coli* rapidly gives a positive reaction.

Tube (c) is split into two. Methyl Red is added to one portion (page 61). *E. coli* produces an acid reaction. To the second portion is added O'Mearas reagent (page 62) to test for the production of acetoin. *E. coli* gives a negative reaction. The colour only develops slowly and the tests should be retained for several hours.

Slope (d) is examined for growth. Growth (utilisation of citrate) produces an alkaline medium and the colour changes from green to bright blue. *E. coli* gives a negative reaction.

The number of tubes giving positive and negative results in test (a) should be scored, and the Most Probable Number of *E. coli* should be calculated from the tables and the appropriate dilution factor.

## 6.3 Faecal Streptococci

Presumptive positives are indicated by growth at 35 °C on Azide Dextrose broth, which also allows the organisms to be counted. The production of red colonies on Slanetz and Bartley's medium at 44 °C is also indicative of faecal streptococci, and this medium is frequently used in membrane filtration systems. These media may be in either the solid or liquid form, but solid media cannot be remelted once prepared. Azide is toxic and the manufacturer's advice for the preparation of these media should be followed. It may also become explosive in the presence of certain metals and should be disposed of carefully. Azide can be destroyed safely by treatment with excess sodium nitrite.

The presumptive faecal streptococci can be confirmed by subculture onto a plate of Bile Aesculin Azide Agar. This contains the water soluble glycoside aesculin, which is hydrolysed by *S. faecalis* and other Lancefield Group D streptococci to glucose and aesculetin. Aesculetin reacts with ferric citrate present in the medium to produce a black/brown colour within a few hours (Chapter 6).

### 6.4 *Clostridium perfringens*

This organism forms highly resistant spores and testing consists of heating the water sample to 75 °C for 10 minutes to kill all vegetative organisms. The water is then dispensed into Double Strength Reinforced Clostridial Medium in screw capped bottles which are topped up with single strength medium to reduce the air space. The method may be made quantitative as well as qualitative by use of the appropriate dilutions. Reinforced Clostridial Agar may also be used for plating out the organism. In both cases a positive result is indicated by growth. The organisms obtained may be subcultured for further tests.

### 6.5 Pseudomonas

Tests for *Pseudomonas aeruginosa* are not used in the standard tests for the examination of water. Tests for this organism are frequently used by manufacturers of food, drinks, and pharmaceuticals. *Pseudomonas* is a Gram-negative rod, giving positive catalase and oxidase tests and producing pigmentation.

Membrane filtration is recommended rather than tube counts, unless the water contains large amounts of particulate matter. The membranes may be incubated on a medium containing cetrimide at a temperature of 42 °C and positive colonies are those that show blue, green or brownish pigmentation. The organism may be confirmed by subculturing onto milk agar containing cetrimide at 42 °C for 24 hours. Incubation at 42 °C inhibits the growth of other pseudomonads which grow at lower temperatures. Growth associated with casein hydrolysis (clearing of medium around colonies) and pigment production is counted as positive for *Pseudomonas aeruginosa.*

### 6.6 Pathogens

Technically the search for pathogens is difficult as they are usually heavily outnumbered by the coliforms and other species which overgrow them on many media. The organisms most likely to be isolated are the Salmonellae with *S. typhi* being a distinct possibility. It is recommended that the search for pathogens of this type should not be attempted unless adequately trained personnel and a laboratory suitable for handling high category pathogens are available. If there is reason to suspect that the samples may contain species belonging to the genera *Salmonella*, *Shigella*, or *Campylobacter*, then liaison with the local Medical Officer of Health is essential.

## 7 INTERPRETATION OF RESULTS

The *routine* samples taken from a water supply should be reviewed regularly and at least once a year.[2] This will give a profile of microbiological quality over a period of time, which makes any change more obvious. A minimum of fifty samples should be used to create the profile. Only routine samples should be used, as the inclusion of any repeat sample taken to investigate a problem will bias the profile. Any complaints or problems arising during the period should be correlated with unusual conditions such as a breakdown in the distribution system or abnormal weather. If *E. coli* or other coliforms are detected then the frequency of sampling should be increased.

The results may be divided into several groups:[2]

(a) Excellent. No coliforms or *E. coli* detected in any 100 ml sample.
(b) Satisfactory. One to three coliforms per 100 ml of sample. No *E. coli* should be present. Coliforms should not be present in consecutive samples and not present in more than 5% of samples.
(c) Suspect. Four to nine coliforms per 100 ml of water. No *E. coli* should be present. Coliforms should not be present in consecutive samples and not present in more than 5% of samples.
(d) Not satisfactory for any one of the following reasons:
(d1) Ten or more coliforms present in any one sample of 100 ml.
(d2) One or more *E. coli* present in any one sample of 100 ml.
(d3) Coliforms present in consecutive samples.
(d4) Coliforms present in more than 5% of samples.

## 8 NON-DRINKING WATER

*Legionella* is a Gram-negative rod isolated in 1976 following an epidemic of pneumonia (known as Legionnaires disease) amongst members of the American Legion. *L. pneumophila* has now been isolated from a wide variety of aqueous systems, including cooling systems, water towers, and items such as shower heads, and should be considered a risk in any system where an aerosol is created. An increasing number of other species of this genus are now being discovered.

The isolation and identification of *Legionella*[5] needs large volumes of water (up to 10 litres). The organisms are concentrated by membrane filtration and cultured onto selective media. After concentration, but before culturing, the samples can be heated to 50 °C for 30 minutes or the pH lowered to 2.2 for 10 minutes to reduce the numbers of other competing bacteria. A number of different media are used, but growth on BCYE agar (buffered charcoal-yeast extract agar) is recommended by Vesey *et al.*[5] All *Legionella* species also require the amino acid L-cysteine for growth and any isolate requiring cysteine is assumed to be *Legionella*. The hydrolysis of hippuric acid to benzoic acid and glycine, followed by the estimation of glycine with ninhydrin, may be used as a spot test for *L. pneumophila* (Figure 9.4).

Several other methods are available for identification, including GLC of the branched chain fatty acids of the cell wall, ELISA, membrane immunoassay and immunofluorescent assay. Identification by serology is beyond the scope of most analytical laboratories and identification to species level is best carried out in a pathology or Public Health Laboratory.

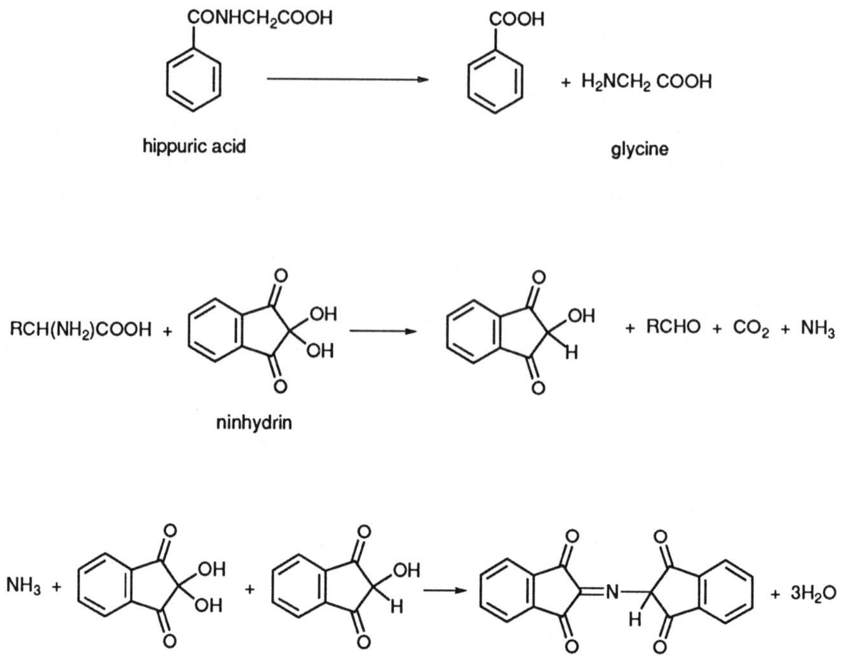

CONHCH$_2$COOH → COOH + H$_2$NCH$_2$ COOH

hippuric acid glycine

RCH(NH$_2$)COOH + [ninhydrin] → + RCHO + CO$_2$ + NH$_3$

ninhydrin

NH$_3$ + → + 3H$_2$O

**Figure 9.4** *Hydrolysis of hippuric acid and estimation of the glycine released by ninhydrin. Ninhydrin reacts quantitatively with amino acids to give a purple colour which can be measured spectrophotometrically*

## 9 REFERENCES

1. B.H. Olson and L.A. Nagy, 'Microbiology of Potable Water', in 'Advances in Applied Microbiology', ed. A.I. Laskin, Vol. 30, Academic Press, 1984, pp. 73–132.

2. 'Reports on Public Health and Medical Subjects, No. 71, The Bacteriological Examination of Drinking Water Supplies'. HMSO, London, 1982.

3. 'American Public Health Association, Standard Methods for the Examination of Water and Wastewater', 17th Edn., Washington, USA 1990.

4. 'Quality of Water Intended for Human Consumption'. EC Directive, 1980, (80/78/EC).

5. G. Vesey, A. Corbin and J. Dennis, 'Identification of *Legionella* Species', in 'Identification Methods in Applied and Environmental Microbiology', ed R.G. Board, D. Jones and F.A. Skinner, Society of Applied Bacteriology, 1992, Vol. 29, pp. 111–125.

CHAPTER 10

# Sterility, Sterility Testing, Disinfectants and Preservatives

## 1 INTRODUCTION

There are a large number of chemicals which are able to kill micro-organisms in addition to antibiotics. Many of these are used on surfaces or skin and go under a variety of names. Chemicals used in the presence of dirt or dense bacterial populations are known as disinfectants. Antiseptics are used on the skin to reduce the bacterial population and thus the risk of infection. Sterilants are compounds used to sterilise enclosed spaces, whilst preservatives are compounds which stop bacteria attacking organic matter. These definitions form a spectrum and compounds frequently overlap from one group to the next.

Chemicals used against micro-organisms may be divided into two groups, bacteriostatic and bactericidal (biocidal). Bacteriostatic compounds are those which stop bacteria growing, whilst bactericidal compounds are those which kill bacteria. The division is not clear cut as many compounds are bacteriostatic at low concentrations and bactericidal at high concentrations. The extent of the bactericidal effect of any compound is governed by four factors. These are:

(a) The concentration of the compound.
(b) The density of the bacteria.
(c) The length of contact time between the compound and bacteria.
(d) The presence of organic matter (dirt).

## 2 STERILISATION

The process of sterilisation is required for any product used in a situation where there is a risk of infection. Sterilisation requires the application of a chemical (biocide) to a product, or the physical removal of micro-organisms from the product. The objective is to kill the bacteria, the definition of bacterial death being the irreversible loss of reproductive ability. There are a number of chemical and physical methods available to achieve this objective. The physical methods

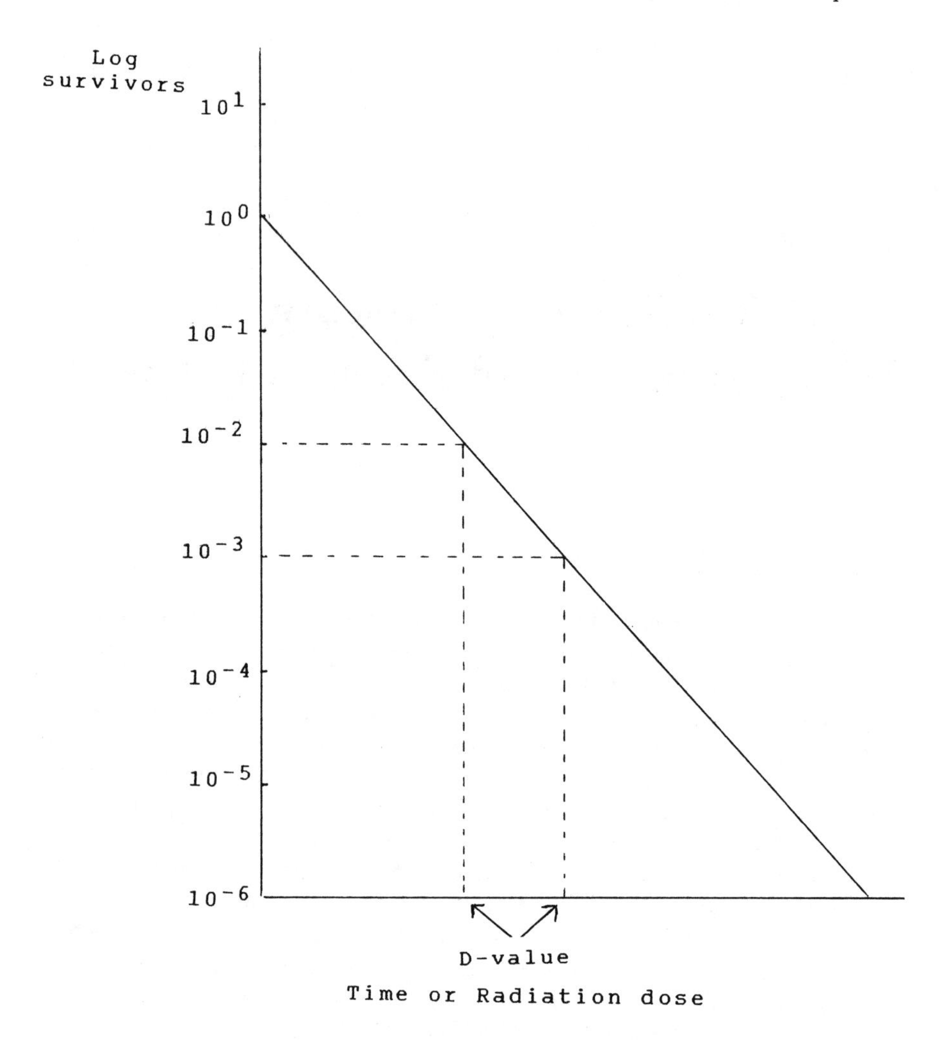


**Figure 10.1** *Plot to show the exponential loss of cell viability during sterilisation* (Adapted with permission from W.B. Hugo and A.D. Russell, 'Pharmaceutical Microbiology', 5th Edn., Blackwell Science Ltd., Oxford, 1992)

include heating, radiation, gas or steam treatment, and filtration. The success of the process depends on a suitable combination of method and time.

In all cases where a product requires sterilising, there is a risk of damage to the product, and it is necessary to balance the risk of damage against the risk of failing to sterilise the product successfully. It is therefore necessary to choose a method that causes maximum destruction/removal of the bacteria, with minimum damage to the product.

Although a knowledge of the microbiological origin and history of the product is useful, sterilisation processes can be developed on a 'worst case' scenario. Bacterial spores are more resistant than vegetative cells and can be used as reference organisms to test the efficiency of the process. The spores used vary depending on the process in operation. *Bacillus stearothermophilus* spores are used for moist heat, *B. subtilis* for dry heat, and *B. pumilus* for radiation.

When exposed to a sterilisation process, cells lose their viability exponentially. This is independent of the number present at time zero. This can be shown logarithmically (Figure 10.1). A linear result is generally obtained, although this may be modified by a slightly reduced kill in either the early or late stages. In a few cases an activation stage may be seen during which spores give an increased count for a short period.

The D-value is used to describe the resistance to a sterilising agent. This is the time taken to produce a 90% reduction in viable cells *i.e.* the time corresponding to the $10^{-1}$ fraction in Figure 1, when using a fixed temperature or radiation dose.

Another term used is the *Z*-value. This measures the effect of temperature changes on the thermal resistance of the organism, and is found by plotting the

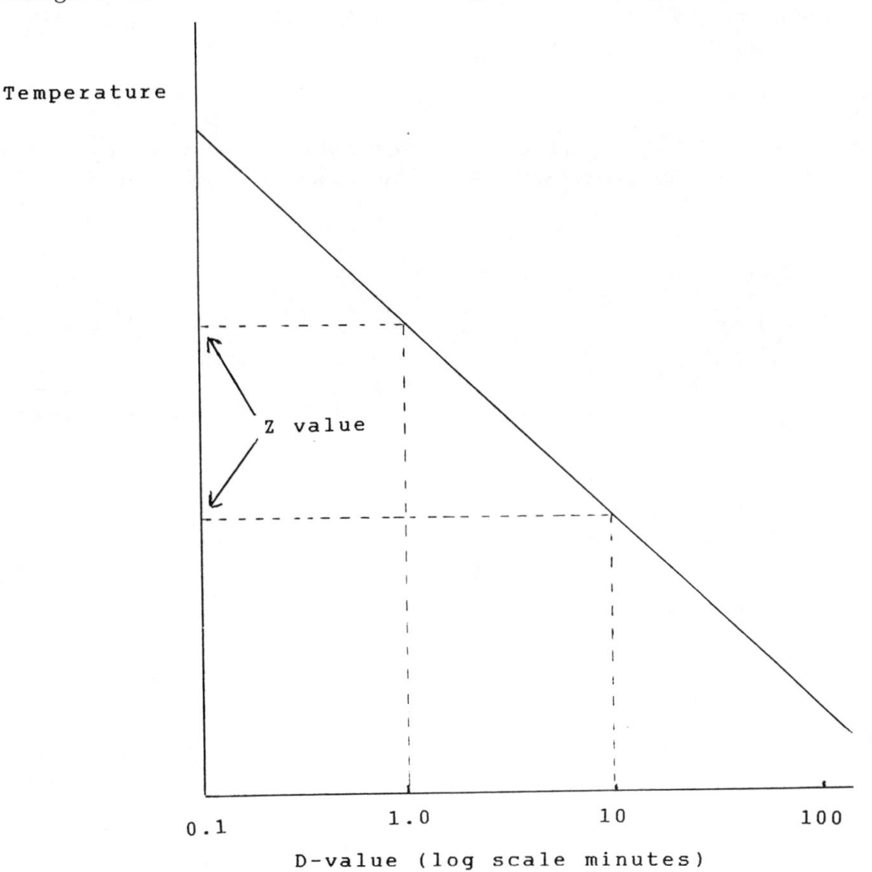


**Figure 10.2** *Plot to show how the Z-value is determined*
(Adapted with permission from W.B. Hugo and A.D. Russell, 'Pharmaceutical Microbiology', 5th Edn., Blackwell Science Ltd., Oxford, 1992)

temperature against the log of the *D*-value. The *Z*-value is then the increase in temperature needed to reduce the *D*-value by 90% (one log cycle) (Figure 10.2).

It is obvious from Figure 10.1 that the removal of viable cells is time dependent and will be influenced by the initial level of contamination. It is also obvious from Figure 10.1 that as the sterilisation process proceeds, the number of cells only approaches zero and does not actually reach it. This is not compatible with the concept of sterility, which is an absolute term, *i.e.* the sample is either sterile or not: there are either micro-organisms present or there are not. This can best be described by pointing out that if there was one contaminant per bottle, then a 90% reduction does not mean $10^{-1}$ organisms per bottle, it means one organism in ten bottles. Therefore, irrespective of the number of log cycles to which a product is subjected, viable organisms will still be found if a sufficiently large sample is taken (Figure10.3).

The usual standard applied for pharmaceutical products is that the probability of a non-sterile unit should be less than 1 in $10^6$ units ($10^{-6}$). The sterilisation procedure necessary to reach this figure can be determined if the *D*-value is known using the inactivation factor (*IF*)

$$IF = 10^{t/D}$$

where *t* is the constant time (heat) or dose (radiation). Therefore if there are 1000 spores ($10^3$) per unit present initially, an inactivation factor of $10^9$ is needed to

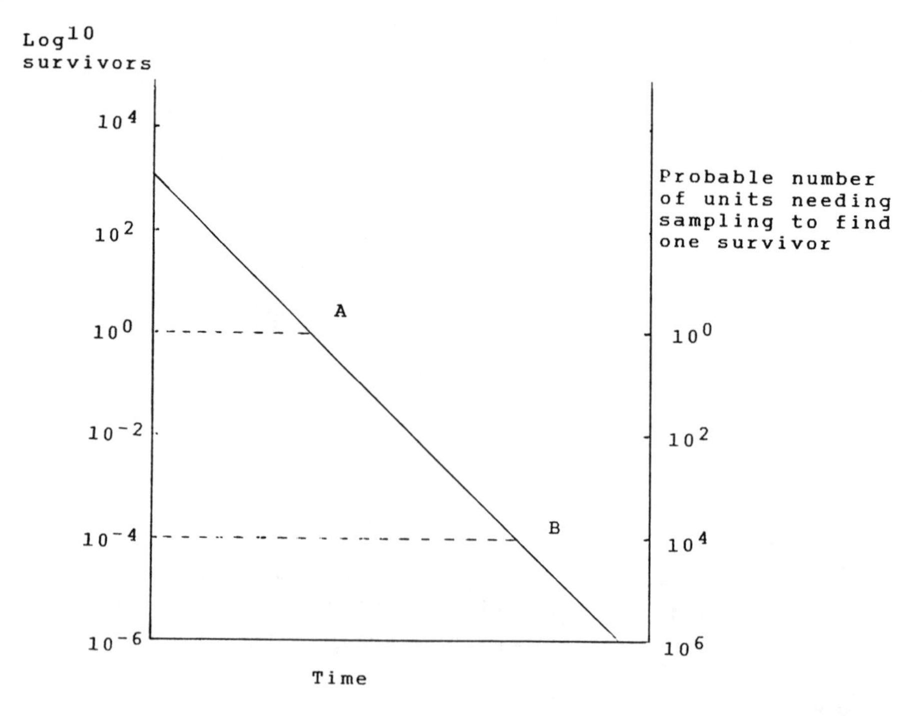


**Figure 10.3** *At point A there is a theoretical probability of one organism per unit. At point B there is a theoretical probability of $10^{-4}$ organisms per unit, or one organism per $10^4$ units*

produce a probability of $10^{-6}$ spores per unit or one spore in $10^6$ units. The sterilisation process therefore requires a 9 log cycle reduction in contamination, or 9 × the *D*-value of the contaminant (Figure 10.3). In practice, the *D*-value of the reference micro-organism is normally used. As these are spores chosen for their resistance to the process under consideration, it is assumed that they will cover a worst case scenario.

Five methods of sterilisation are normally recognised. These are dry heat, moist heat (autoclave), filtration, gas treatment and radiation. Heat is widely used and is very efficient, but can obviously only be used for thermostable products. Heat used at high humidity is more effective, and is used in the range 121–134 °C, whereas dry heat needs a temperature of 160–180 °C. The use of moist heat in an autoclave is generally corrected to 121 °C to allow for heating and cooling cycles, and a commonly used time is 20 minutes, which is adequate for most purposes. Probes are available to ensure that an adequate temperature is reached in the centre of the sample. Steam is also widely used for sterilisation, especially for the sterilising of pipes and large scale fermentation equipment.

Dry heat is generally used at 180 °C for a period of two hours for the sterilisation of glassware, metal instruments and thermostable products. Spore strips of *Bacillus stearothermophilus* may be placed in the oven as a biological sensor, and removed for culturing at the end of the heating period. Growth indicates that the sterilisation process was inadequate, whilst no growth shows a suitable temperature was maintained for the requisite time.

Sterilisation by the use of gases usually involves either ethylene oxide or formaldehyde. Both are broad spectrum biocides but neither is as effective as heat. Both may be used for temperature sensitive systems. Both must be used under the correct conditions as they are toxic, and ethylene oxide is explosive when mixed with air. Ethylene oxide is highly penetrative, but formaldehyde is not and is generally used where surface sterility is required.

Sterilisation by radiation is being used increasingly. Several types of source, such as high energy gamma rays or UV light, may be used for sterilisation of heat sensitive materials, such as dry pharmaceutical products. The main target is the DNA of the cell, and cells vary widely in their ability to repair radiation damage. The *D*-value (radiation dose) required may be established using the spores of *B. pumilus* as standard. Ultra-violet light is also widely used to reduce the bacterial count in items such as glove boxes and aseptic cabinets.

Sterilisation by filtration differs from the above methods as it removes cells rather than destroys them. It also removes particles and can be used for clarification, carrying out both functions simultaneously. Filtration can be used for both heat sensitive liquids and air supplies to aseptic areas.

There are several physical mechanisms for the removal of cells and particles. These include sieving, adsorption, and trapping in the matrix. Membrane filters work mainly by sieving, whilst filters such as fibrous pads and sintered glass are depth filters, and work by adsorption and entrapment. Depth filters can suffer from microbial multiplication within the filter, which causes contamination in the filtrate, a problem known as grow through.

The pore size of membrane filters ranges from 0.2–0.45 $\mu$m, with the ones at the lower end being generally used. Membrane filters are adequate for removal of bacteria, but they do not remove all viruses. They are frequently used in series after a sintered glass filter, which removes coarse particles, and reduces the probability of the membrane filter becoming clogged.

## 3 STERILITY TESTING

Sterility testing assesses whether a product (usually medical or pharmaceutical) contains any contaminating micro-organisms. It is carried out by incubating the product (either whole or in part), in a nutrient medium allowing bacterial and fungal growth. It is therefore destructive testing, and as it is obvious that the whole product will not be used, the question of random sampling and statistically significant sampling arises.

Following any sterilisation process, any micro-organism surviving will be stressed, and will need to be given every opportunity to grow. The medium chosen and growth conditions therefore need to be optimal. Accidental contamination must be avoided at all costs, and product testing should be carried out under aseptic conditions by a fully competent member of staff.

Standard sampling and test procedures are given in the BP[1] and must be followed for any product carrying the BP label. These tests are also suitable for many other products. The method of assessing the sample will obviously vary depending upon the nature of the product, but the following methods are widely used. The sample may be inoculated directly into a suitable nutrient medium. Media recommended by the BP are sodium thioglycollate for the growth of anaerobes, and soya bean casein digest medium for the growth of aerobic bacteria and fungi. The growth temperature chosen will vary depending on the organism involved in the test.

Liquid products and water soluble solids may be tested by membrane filtration. Solids would first be dissolved in a suitable diluent, such as 1/4 strength Ringers solution. After the liquid has been passed through the sterile filter (pore size $< 0.45\ \mu m$), the filter is cut up aseptically, and pieces are placed into an appropriate culture medium for incubation. Liquid products may also be tested by adding a concentrated preparation of culture medium to a container with the product *in situ*. The concentration of medium added is such that the combination of medium and product gives single strength medium. The whole container is then incubated at the appropriate temperature.

One problem that may arise is when an antibiotic or preservative is the product, or part of the product under test. When this happens, the material must be inactivated or removed before sterility testing can take place. There are several methods of achieving this. Antibiotics, such as the penicillins, may be inactivated by the addition of the enzyme β-lactamase, whilst the action of sulphonamides can be blocked by the addition of *p*-aminobenzoic acid. Products containing preservatives or antimicrobial agents, such as benzoic acid, alcohols, or phenols, are diluted to the level at which the compound becomes ineffective. Products containing quaternary ammonium compounds can be inactivated by the addition of Tween, whilst many compounds containing heavy metals can be inactivated by the use of thiols. Products containing inhibitory agents for which there is no inactivating agent, should be passed through a membrane filter, which is then washed well and transferred to an appropriate culture medium.

Positive controls must be used in sterility testing. These are designed to show that the organism is capable of growing under test conditions. The test organisms used must be ones that could be reasonably expected to be found in a non-sterile batch of the product. Also, if the product contains an antibiotic or preservative, then the control organism should be sensitive to it. Growth of the control organism will then show that the antibiotic or preservative has been satisfactorily inactivated or diluted.

Organisms suggested as positive controls by the BP are *Staphylococcus aureus* (a typical aerobic micro-organism), *Bacillus subtilis* (aerobic spore former), *Clostridium sporogenes* (anaerobic spore former), and *Candida albicans* (yeast and fungus).

## 4 SAMPLING

The sampling of a number of items is discussed in the chapters on food, water and counting. It is obvious from the fact that sterility testing is destructive, that one cannot test the whole batch, and therefore sampling on a statistically significant basis must be carried out. The statistics are discussed briefly with a worked example,[2] and also in the BP.[1]

A small number of simple examples are probably the best way of demonstrating the situation.

### Example 1

Let $x$ be the proportion of contaminated tins in a batch. Let $y$ be the proportion of non-contaminated tins in the same batch. Then $x + y = 1$.
Consider a large batch of tins of which 20% are contaminated.
The probability of a randomly selected tin being contaminated is $20\% = 0.2 = x$.
The probability of a tin being non-contaminated $= y = (1-x) = 0.8 = 80\%$.
If a second random tin is taken, then the probability of this being contaminated is also 0.2.
The probability of both tins being contaminated is $0.2^2 = 0.04 = 4\%$.
The probability of both tins being non-contaminated is $0.8^2 = 0.64 = 64\%$.
The probability of one contaminated and one non-contaminated tin being chosen is $1 - (x^2 + y^2) =$ which is $1 - (0.64 + 0.04) = 0.32 = 32\% = 2xy$.

Thus when one tin in every five is contaminated, there is only a 36% (*i.e.* 0.32 plus 0.04) chance of detecting the contamination.

### Example 2

With decreasing levels of contamination there is an increase in the probability of selecting two non-contaminated samples as shown below.

(a) Level of contamination 20% (two hundred tins per thousand).

Probability of obtaining two non-contaminated samples

$0.8^2 = 0.64 = 64\%$

(b) Level of contamination 2.0% (twenty tins per thousand).

Probability of obtaining two non-contaminated samples

$0.98^2 = 0.9604 = 96.04\%$

(c) Level of contamination 0.2% (two tins per thousand).

Probability of obtaining two non-contaminated samples

$0.998^2 = 0.996 = 99.6\%$

Thus with much lower levels of contamination it becomes much more likely that two non-contaminated tins will be selected.

### Example 3

The results obtained for different numbers of samples for a given value of $x$ (proportion of contaminated tins in batch), show that sterility testing does not detect low levels of contamination reliably. For example, with a contamination level of 0.2% (*i.e.* two cans in $10^3$), which is several orders of magnitude greater than the allowable contamination rate for pharmaceutical products (< one in $10^6$) the following probabilities are obtained:

Probability of obtaining one non-contaminated sample

$0.998 = 99.8\%$

Probability of obtaining two non-contaminated samples

$0.998^2 = 0.996 = 99.6\%$

Probability of obtaining three non-contaminated samples

$0.998^3 = 0.994 = 99.4\%$

Probability of obtaining four non-contaminated samples

$0.998^4 = 0.992 = 99.2\%$.

However, the results show that if two tins are taken at random for testing there is a probability of 99.6% that the contamination would not be detected, whilst if the number of samples rises to four the probability of non-detection is still over 99%. The results also show that if varying sample sizes are used (for a given level of contamination), then as the sample size increases, the probability of the batch being passed as sterile falls. It is therefore important to point out that, as sterility testing only tests samples from a batch, the conclusion is not that the batch is sterile, but that the batch has passed the sterility test. There is obviously an important difference between the two.

## 5 DISINFECTANT TESTING

There are a number of tests designed to determine the effect of disinfectants on bacterial populations. Two tests which have been used widely in the past are the Rideal–Walker and Chick–Martin tests for determining the phenol coefficient of disinfectants in a hospital situation. These tests compare the efficiency of the disinfectant with phenol. They should only be used to assess a phenol based disinfectant and have been widely criticised as inappropriate and non-reproducible.[2,3] Details of these tests can be found in British Standards if required.[4,5]

Minimum Inhibitory Concentration (MIC) tests give a rough estimate of the effectiveness of antibiotics and are used widely in hospitals and the pharmaceutical industry. They have also been criticised as the time of exposure during the test is much greater than the 'in use' exposure time, the test temperature (usually 37 °C) is higher than the 'in use' temperature, and there is insufficient organic material present to make the test realistic.[3] In spite of these criticisms MIC tests are still widely used. They are usually done as a tube assay, but may also be done using agar.

The technique, which is quick and simple, consists of placing a known quantity of the antibiotic or preservative in a known volume of nutrient broth. A geometric series of dilutions is prepared as follows. The antibiotic is measured into 10 ml of nutrient broth in tube 1 and mixed well. Five ml are removed and placed into tube 2, which contains 5 ml of sterile broth. This is mixed and 5 ml are removed from tube 2 and added to tube 3 which also contains 5 ml of sterile broth. This is mixed and the procedure repeated for tube 4 *etc.* The final 5 ml removed from the last tube is discarded. Each tube is inoculated with one drop of the bacterial culture being used to challenge the antibiotic, and incubated, usually overnight. The tubes are assessed visually (turbidity/no turbidity) for growth/no growth. The clear tube (*i.e.* no growth), with the most dilute antibiotic preparation, is taken as the practical MIC value. Technically however, the true MIC value lies between the concentrations of antibiotic in the last tube showing growth and the first tube showing no growth.

The Rideal Walker and Chick Martin tests have been replaced by the Kelsey–Sykes test[6] which assesses the disinfectant at its 'in use' concentration under both clean and dirty conditions. There is a considerable amount of preparation involved and the tests are probably best carried out in specialist laboratories.

The test consists of adding microbial cultures to disinfectants at their 'in use' concentration and removing samples at prescribed time intervals. These samples are either plated out onto nutrient agar or added to a liquid recovery medium. The plates or tubes are incubated and checked for growth. Full details can be found in the original reference[6] or in the British Standard specification.[7]

## 6 REFERENCES

1. 'British Pharmacopoeia', British Pharmacopoeia Commission, HMSO, London, 1980.
2. W.B. Hugo and A.D. Russell, 'Pharmaceutical Microbiology', 5th Edn., Blackwell Scientific Publications, Oxford, 1992.
3. 'Mackie and McCarteney, Practical Medical Microbiology', 13th Edn., ed. J.G. Collee, J.P. Duguid, A.G. Fraser and B.P. Marmion, Churchill Livingstone, Edinburgh, London and New York, 1989.
4. British Standard 541, Method for Determination of the Rideal–Walker Coefficient of Disinfectants, 1985.

5. British Standard 808, Method for Assessing the Efficacy of Disinfectants by the modified Chick–Martin test, 1986.
6. J.C. Kelsey and G. Sykes, A New Test for the Assessment of Disinfectants with Particular Reference to Their Use in Hospitals, *Pharmaceutical Journal*, 31st May, 1969.
7. British Standard 6905, Method for the Estimation of Concentration of Disinfectants used in 'Dirty' Conditions in Hospitals by the Modified Kelsey–Sykes test, 1987.

CHAPTER 11

# Microbiological Assay

## 1 INTRODUCTION

Microbiological assay is a technique in which the potency or concentration of a compound is assessed by determining its effect on micro-organisms. The principles are discussed by Roberts and Boyce.[1] Microbiological assay is a legal QC requirement for the assay of a number of antibiotics, in both the British Pharmacopoeia (BP) and United States Pharmacopoeia (USP).

Bioassay compares a reference standard and an unknown sample, the two preparations being measured simultaneously.

**It is essential that both reference and sample are of the same type.**

Some known formulation of the active compound is designated as the standard preparation, and test samples can be compared with it. A quantitative value for the relative potencies of the standard and test samples is thus obtained. When the test and sample preparations are compared, the less potent of the two should behave as if it were a dilution of the more potent in some inert diluent. This is the ideal situation, and is not always true for materials containing more than one active ingredient. Hewitt and Vincent[2] give a list of standards.

The variable response is a major problem in bioassay, and historically many unrecognised sources of error occurred, producing unreliable results. As fundamental principles have become better understood, assays have become considerably more reliable in competent hands.

Random variation is found in biological assays, especially macrobiological assays, due to differences between individuals. This problem does not arise in microbiological assays as large numbers of micro-organisms are used. For example, a 1 ml inoculum may contain $10^8$ bacteria. Random variation in microbiological assays is therefore in the methodology rather than the individual organism, and can be minimised by the correct application of theory and technique. However, complex statistical analysis of results does not compensate for poor experimental design.

Bioassays of antibiotics and vitamins have many common features, but there are differences. The most basic difference is that antibiotics and preservatives inhibit growth, whilst vitamins and amino acids enhance growth.

## 2 SAMPLING

Sampling homogeneous liquids causes few problems and this is also the case for heterogeneous liquids as long as they are homogenised adequately. Sampling heterogeneous solids is more problematical. When sampling tablets, at least twenty tablets should be weighed, ground, and mixed well. An aliquot is then weighed and used for assay. Antibiotics are frequently present in very low concentrations and careful weighing and measuring of volumetric solutions is essential.[2] Creams and similar emulsified materials should be thoroughly mixed before weighing and detailed methods of preparation are found in the BP.[3]

## 3 MICROBIOLOGICAL ASSAYS

A lot of preliminary work is necessary for microbiological assays, and if only a small number of samples are expected irregularly, the method is inefficient. Two assay methods are normally used, agar diffusion and tube assays. They have several common features:

(a) The compound being assayed must influence the growth of the test organism.
(b) A varying response in growth must be produced by addition of varying quantities of the test material.
(c) The growth medium must contain an excess of all the compounds required by the test organism for growth. The exception to this is the compound being assayed which should be totally absent from the basic medium.
(d) The assumption is made that the compound being assayed is the only growth promoting or inhibiting compound present.

### 3.1 Agar Diffusion Assays

These are usually carried out on plates. The assay system is identical for antibiotics and vitamins. A liquid agar (50 °C) preparation of medium minus the compound to be assayed is inoculated uniformly with a sensitive organism. The agar is poured into a petri dish and is allowed to solidify. The assay material and suitable reference standards are placed in individual reservoirs. These diffuse out into the agar, and the organism grows to form zones after a suitable period of incubation. The zone is one of inhibition (no growth) when antibiotics or preservatives are used. The zone is one of growth or exhibition when vitamins or amino acids are used.

**The width of the zone, which depends on the concentration of active compound used, forms the quantitative basis of the test.** It is assumed that the compound being assayed can diffuse freely throughout the medium being used. This may not always be true, especially at low concentrations.

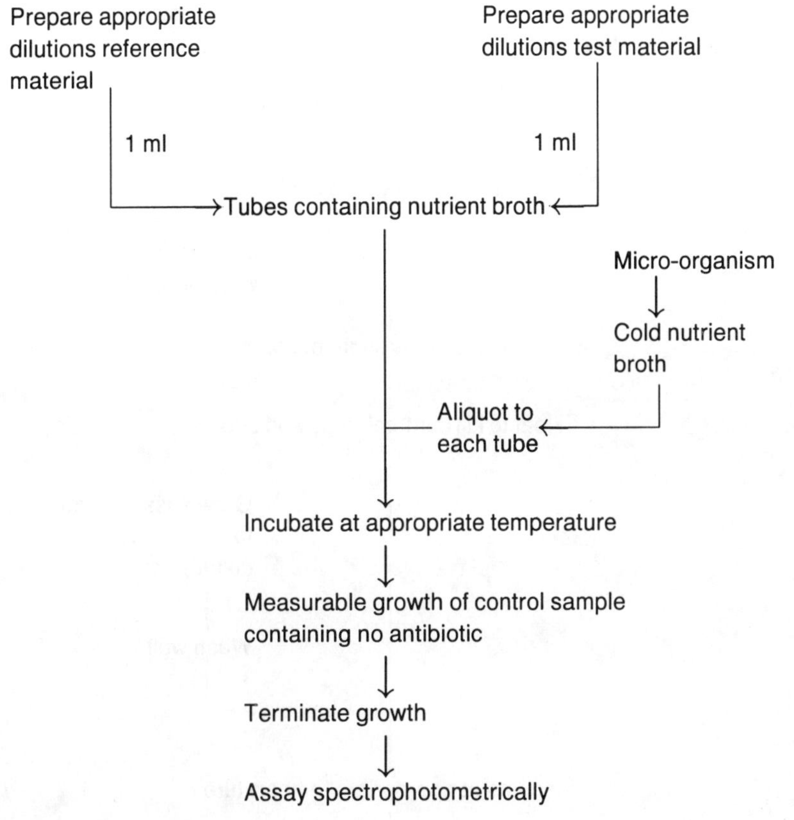


**Figure 11.1** *Tube assay of antibiotics*

## 3.2 Tube Assays

Tube assays use liquid medium, and growth inhibiting and growth promoting systems differ.

*3.2.1 Antibiotics (Growth inhibiting).* A series of dilutions is prepared for reference and test samples (Figure 11.1). Each dilution (1 ml) is added to each of a series of tubes containing nutrient broth. The appropriate micro-organism in cold nutrient broth is added to each tube, and the tubes are incubated. (The medium is chilled to prevent microbial growth prior to the incubation period, which would distort the results). Incubation continues until a control tube which contains no antibiotic gives an adequate response. Bacteria grow over a range of temperatures with an optimum varying from species to species. In the case of bacteria used in bioassays, the optimum is normally in the range of 30–37 °C. Growth is terminated usually by heating to 80–85 °C, although the BP recommends the addition of formaldehyde. Growth in each tube is measured, usually spectrophotometrically at 540 nm or a similar wavelength. **The difference in growth between the reference and test samples is compared and is the basis for calculating potency.**

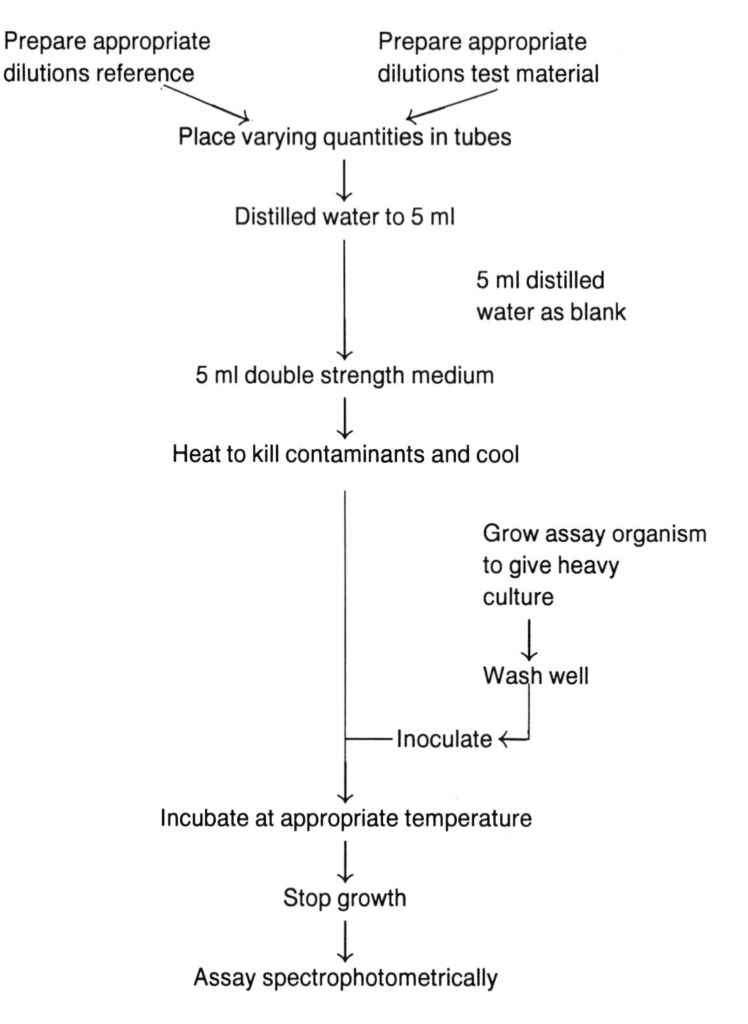


**Figure 11.2** *Tube assay of vitamins*

*3.2.2 Vitamins (Growth promoting).* Reference and test materials are prepared as a series of dilutions (Figure 11.2). Varying quantities of these dilutions (*e.g.* 1, 2, 3 ml) are added to a series of tubes. Distilled water is added to each tube to give a volume of 5 ml. A number of tubes containing distilled water only are also prepared. These act as blanks, some of which are inoculated and others of which are not. A nutrient minimal medium (5 ml) of double the normal strength is added to each tube, giving a total of 10 ml of normal strength medium in each tube. The nutrient medium contains all growth requirements with the exception of the compound being assayed. Several tubes containing a high dose of reference material are used to check that growth will proceed adequately in the presence of the vitamin. The assay tubes are heat treated to kill any contaminating micro-organisms; this assumes that the compound being assayed is heat stable. The

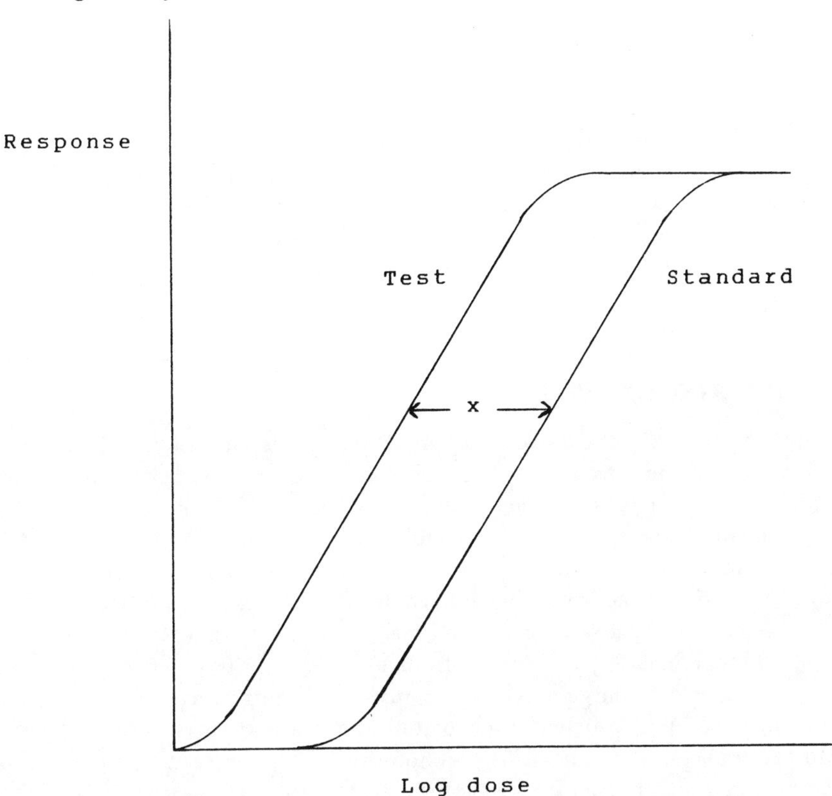


**Figure 11.3** *The dose–response is sigmoidal for both test and standard. The distance (x) measures the efficiency (log potency ratio) of the test relative to the standard. If the two lines coincide then the potency is identical*

tubes are cooled, inoculated with the test organism, and incubated. Growth is stopped and measured after a suitable period of time (usually overnight). **Potency is calculated by comparing the difference between reference and test.**

## 4 DOSE–RESPONSE

The dose–response curve for both test and standard is sigmoidal (Figure 11.3). The distance between the two curves ($x$) measures the effectiveness of the test material relative to the standard. This is the log potency ratio. The two lines coincide if standard and test materials are identical.

## 5 REFERENCE STANDARDS

Microbiological assays compare unknown and reference materials using an assay organism. Ideally all reference standards for a single compound should be identical. This is not practical over a large number of laboratories, and standards are used whose potency can be defined in terms of a single accepted reference compound. Details can be obtained from the World Health Organisation (WHO).

Two types of standard are found, compounds with some established physical/chemical characteristic, and materials for which there is none. The only method of establishing a working standard for the second type is by comparison with the appropriate International Standard. This process is time consuming and technically demanding, and is only worth doing if a large number of samples are expected.

## 6 BACTERIAL GROWTH

There are a number of chemical and physical requirements for bacterial growth, such as carbon and nitrogen sources, pH, and temperature (see Chapter 1). Some antibiotic assay organisms, such as *Bacillus subtilis* and *Escherichia coli*, grow on simple media, others, such as the lactobacilli used for vitamin assay, need an enriched medium containing a wide range of nutritional supplements. Many media are commercially available in the dehydrated form from various suppliers.

The generation time (see Chapter 1) may be extremely rapid for some bacteria under optimal conditions, with cells dividing every 20 minutes. Time may become a problem if large numbers of samples are being processed, as there may be a considerable gap between inoculating the first and last samples. This may lead to significant differences in incubation time. The problem is less serious if non-bacterial organisms, such as *Euglena* or yeasts are being used, as these have a significantly longer generation time. This problem is dealt with by inoculating samples at fixed time intervals and assaying at the same time intervals.

## 7 CULTURES FOR BIOASSAY

Micro-organisms are frequently supplied as freeze dried cultures *in vacuo* in sealed ampoules. Instructions for dealing with these are given in Chapter 3.

Bacteria used for assays can be divided into two groups, spore formers and non-spore formers. The genus *Bacillus*, which are spore formers, are the most commonly used organisms for microbiological assay.

There are advantages to using sporulating as opposed to non-sporulating species. It is simple to produce a standard spore suspension which can be kept for a long time (months or years) at 4 °C. This removes the problems of culture maintenance and regular subculturing. Assays take less time, as the spore suspension is ready immediately, without the necessity of preparing a culture the previous day. Spore suspensions can be prepared by growing *Bacillus* on a sporulation medium and washing the spores off with sterile saline or Ringers solution.

Assays using non-sporulating species (*Escherichia coli* or *Micrococcus lutea*) require cultures to be prepared before they are needed for the assay. This is done by growing the organism on an agar slope overnight, and washing it off with sterile

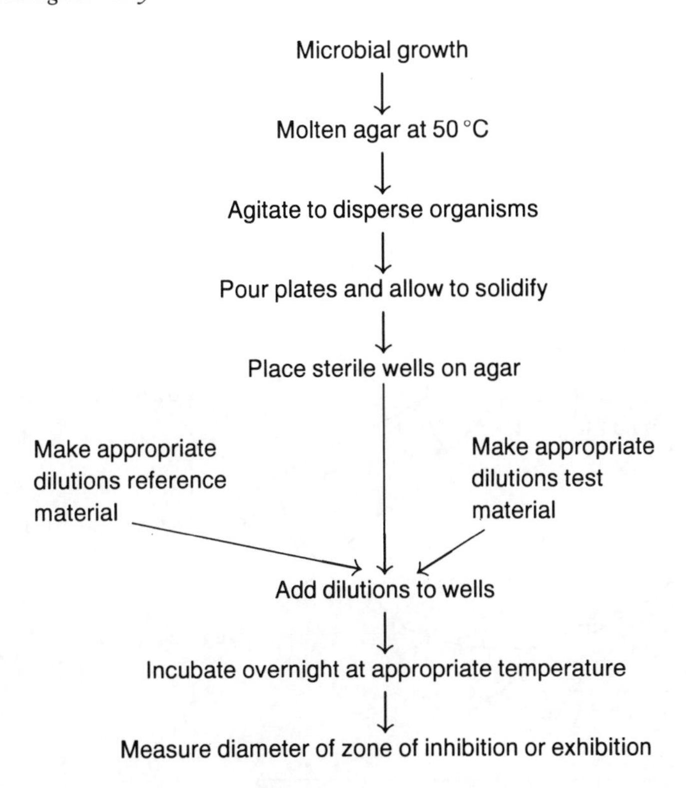


**Figure 11.4** *Procedure for agar diffusion*

saline to give a test suspension. This can be stored in a refrigerator for up to six weeks depending upon the species.

## 8 STANDARDISATION OF THE MICROBIAL INOCULUM

The microbial inoculum for agar diffusion must be standardised before use. Aliquots of the micro-organism are added to molten assay agar at 50 °C, to give a final volume of 10 ml. Higher temperatures than 50 °C will kill some bacteria, lower temperatures will cause lumps to form as the agar is poured into the petri dish. Bacteria are dispersed evenly by rolling the tubes rapidly between the palms of the hands (avoids bubbles). The agar is poured rapidly into sterile petri dishes, placed on a flat surface and allowed to solidify. Bubbles forming in the agar can be removed by passing a bunsen flame rapidly across the surface of the plate before it solidifies.

The growth promoting or inhibiting compound is added (see later). Various concentrations should be used and the experiment should be done in triplicate for each concentration. Plates should be incubated overnight at 30 or 37 °C (see Figure 11.4 for an outline of the procedure). After incubation, the plates are examined for zones of inhibition or exhibition. There should be a clearly defined zone (ideally 15–25 mm) allowing the diameter to be measured (Figure 11.5).

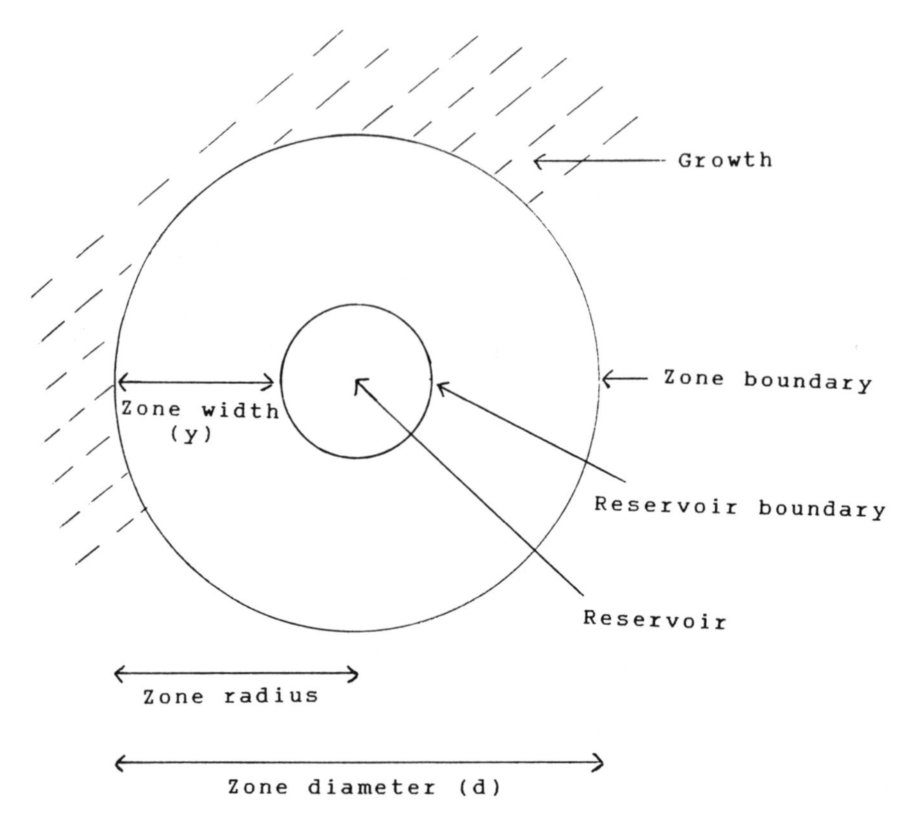


**Figure 11.5** *Diagram showing a zone*

The diameter of the zone is plotted against log of concentration. This indicates the concentration of standards to use in future assays. If the microbial growth is heavy and/or zones are too small, the dilution of the inoculum and/or the concentration of the growth repressor/activator may be adjusted. Suitable parameters should be noted for future reference.

The concentration of bacterial inoculum giving suitable results may be determined using direct microscopic count, viable counts, turbidity tubes, or spectrophotometric methods. These methods may not give the same result so one standard method should be used (see Chapter 5). Spectrophotometric methods may produce varying results between species due to differences in the size of the cells and hence their light scattering properties. Each species should therefore be checked.

## 9 METHODS OF ANTIBIOTIC ASSAY

Antibiotic assays can be carried out in the form of agar diffusion assays, or as tube assays in liquid media.[2,4,5]

## 9.1 Agar Diffusion

Pour plates of susceptible micro-organisms are placed in contact with a reservoir containing the antibiotic. There are several methods of applying the antibiotic to plates. Reservoirs can be cut into agar plates using some instrument such as a sterile cork borer. A number of instruments are commercially available for the automated cutting of reservoirs. The reservoir should be a minimum of 8 mm diameter, as smaller reservoirs are unable to hold a sufficient volume. More commonly, sterile stainless steel cups are pushed into the agar. The reservoir or cup is filled with antibiotic. Alternatively, filter paper discs or unglazed porcelain beads are soaked with antibiotic solution and placed on the plates. The size of the assay discs is critical, and a number of manufacturers sell a standard disc. Paper discs impregnated with antibiotic are widely used in hospitals in the form of multidiscs. These show which antibiotic is effective, and the level of its efficacy.

Antibiotic solutions placed in reservoirs need sterilising. Autoclaving can only be used for a small number of antibiotics, and many need to be sterilised by membrane filtration. The use of cups to hold the antibiotic has the advantage that non-sterile solutions may be used.

The antibiotic should be allowed to diffuse for a short period after application to the plate. The longer it is allowed to diffuse, the larger the zone of inhibition becomes and the steeper the dose-response curve. If the diffusion is carried out at room temperature, growth may occur during the diffusion period, leading to an indeterminate zone. The pre-incubation diffusion should therefore be carried out in a refrigerator to minimise growth. The reduction in temperature reduces the diffusion rate, and the best conditions must be determined by trial and error. A diffusion period of several hours at 4 °C is usually adequate.

Theoretically all doses of antibiotic should be applied simultaneously. When large plates with a number of reservoirs are used, a significant time may elapse between filling the first and last reservoirs. This influences the results by altering the time of diffusion. This problem can be reduced by using an appropriate pattern of sample distribution based on a Latin square or quasi-Latin square.

A Latin square allows a large number of samples to be applied to a square plate in a pre-determined manner. Samples are arranged in a square *e.g.* $4 \times 4$, $6 \times 6$, or $8 \times 8$. In a true Latin square, each antibiotic application appears once in each row and once in each column. A $4 \times 4$ (2+2) plate would have sixteen applications to the plate at two levels for the standard and two for the test. Low and high dose levels would be in the ratio 1:2. Each low standard would appear four times, each high standard four times, each low sample four times, and each high sample four times.

A $6 \times 6$ (3+3) assay may be used for the assay of one standard and one sample at three levels. In this case the dose levels would be a geometric progression in the ratio of 1:2:4. This is used for checking the dose-response of new assays for linearity and parallelism.

The $8 \times 8$ (2+2) system is used widely for the regular assay of one standard and three samples at two levels (1:2). An $8 \times 8$ (4+4) also allows the use of one

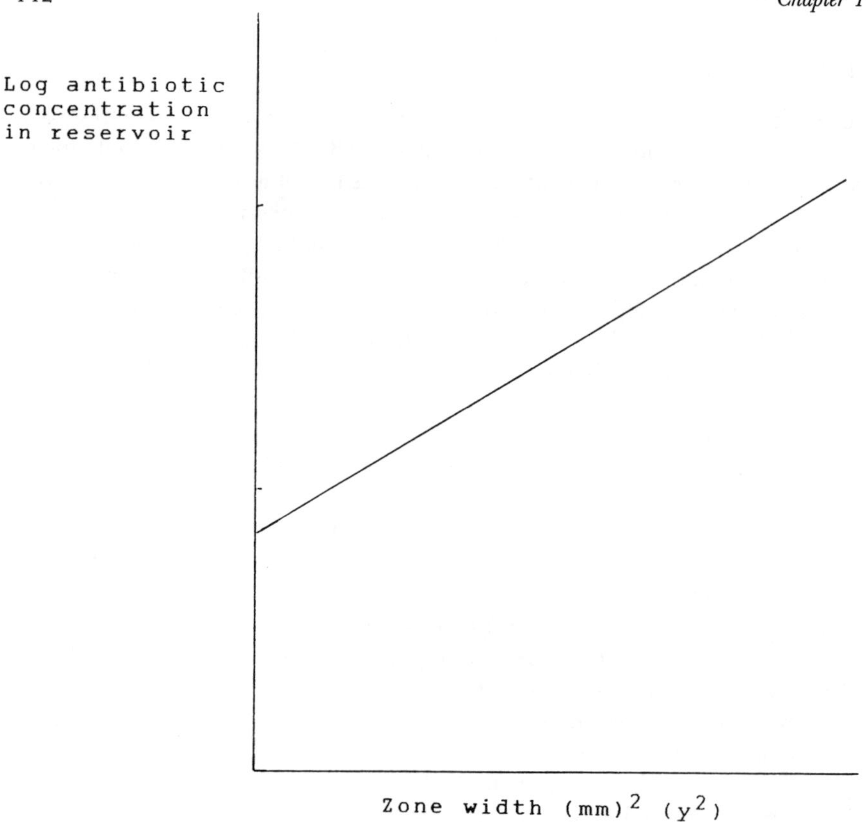


**Figure 11.6** *Intercept on the y axis (x=0) is the minimum concentration of antibiotic needed to form an inhibition zone.*

standard and one test at four levels (1:2:4:8), and is used in very high precision work.

Quasi-Latin squares are considered as a reasonable compromise when a large number of assays is required, *i.e.* when use of a true Latin square would cause an unreasonable work load. An 8 × 8 (2+2) quasi-Latin square allows the assay of six samples and two standards at two dose levels (1:2). Each application appears in four of the eight rows, and four of the eight columns. It therefore appears only four times, compared with eight times on a true Latin square. The use of Latin squares and quasi-Latin squares is discussed in detail by Simpson.[6]

Round plates are also used; these allow four or six applications, each appearing once. Replicate samples are carried out on separate plates, which may cause a problem if they are treated differently.

## 9.2 Factors Influencing Zone Size

The multiplication of micro-organisms will continue until they are inhibited by antibiotic. A growth/no growth boundary is produced which is influenced by a number of practical factors. A nutritionally rich medium produces more rapid growth, forming a smaller zone and a less steep dose-response slope. The size of the inoculum also influences the position of the zone boundary. A heavy inoculum produces a sharper boundary, but gives a smaller zone of inhibition and less steep dose-response line. A light inoculum gives a diffuse boundary, a large inhibition zone, and steep dose-response. The size of the inoculum is therefore a compromise. The time and temperature of diffusion have already been discussed.

Below a certain minimum concentration of antibiotic, no inhibition zone will be formed. This can be shown by plotting the log of the antibiotic concentration against the square of the zone width, $y^2$ (Figure 11.6). The theory of inhibition zones is discussed by Cooper. [4,7]

## 9.3 Dose–Response

The steeper the dose–response curve within certain limits, the more accurate the measurements. The dose–response curve is sigmoidal (Figure 11.3), with the most useful region for assays being the central portion. A calibration curve for the dose–response can be plotted graphically by using reference standards at a series

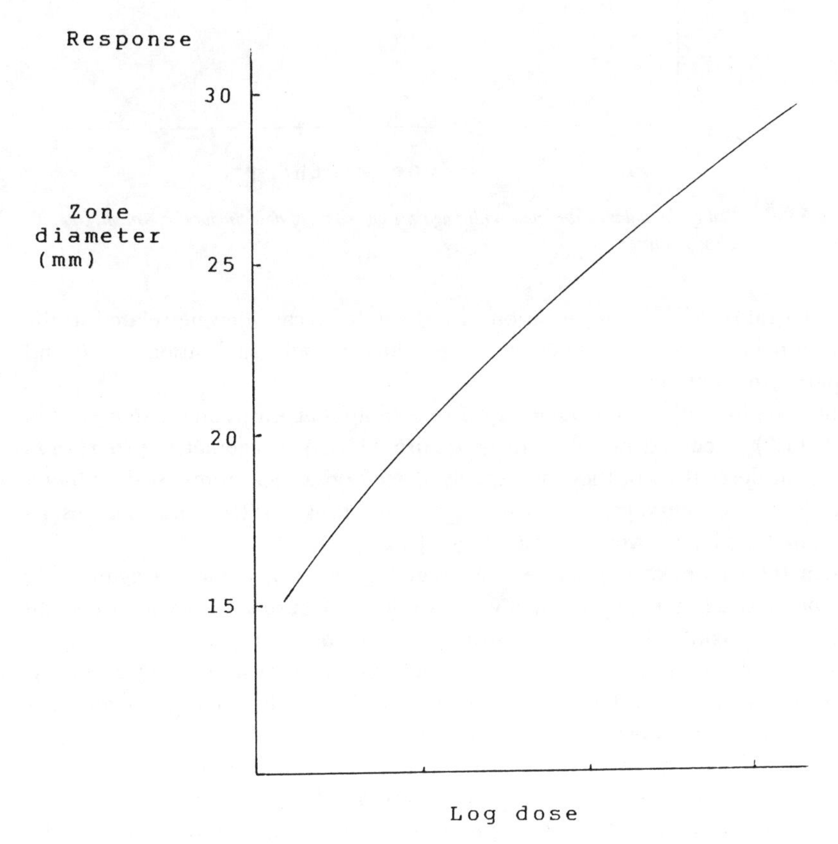


**Figure 11.7** *Plot of logarithm of dose against zone diameter gives a curved line*

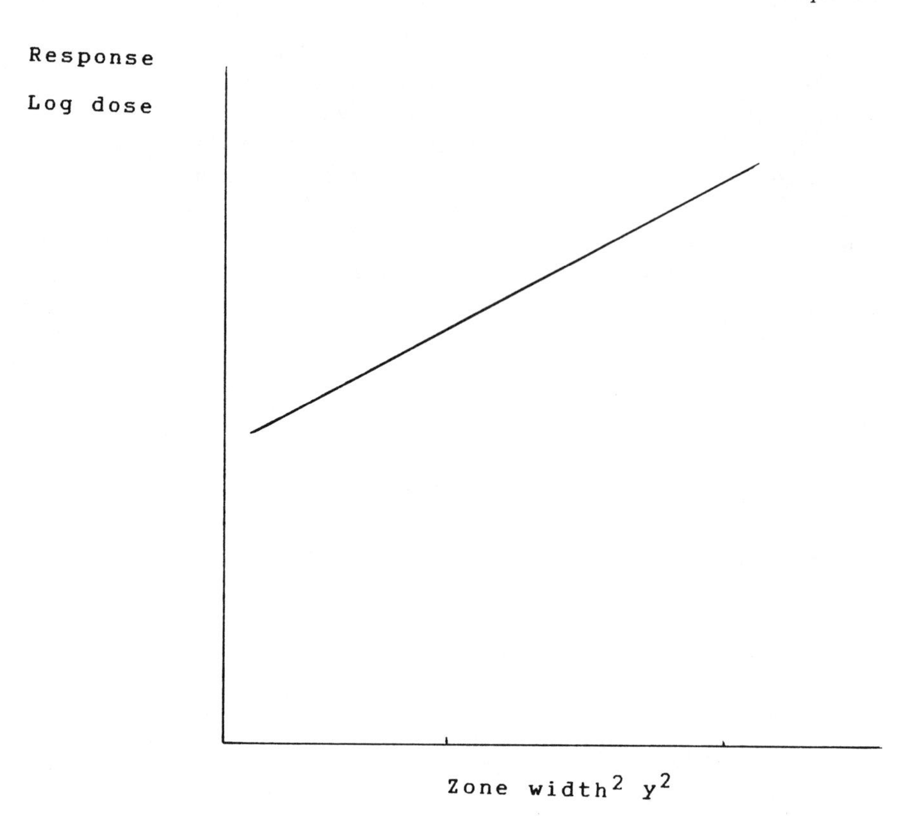


**Figure 11.8** *Plot of the square of the zone width against logarithm of dose produces a straight line when y is large*

of concentrations. The concentration of test solutions can then be related to the calibration curve. It is inconvenient to produce a daily calibration curve and alternative methods are used.

When the logarithm of a range of doses is plotted against zone diameter ($d$ in Figure 11.5), a curved line is formed (Figure 11.7). If a geometric progression (1:2:4:8) of doses is used, they are equally spaced on a logarithmic scale. Only a slight degree of curvature is found over the range 1:2:4, and the results approximate to linear over the dose range of 1:2.[2]

When the square of the zone width ($y$ in Figure 11.5) is plotted against log dose, then a straight line is obtained when $y$ is large (Figure 11.8). When $y$ is small (less than approximately 3 mm), $y$ is proportional to log dose.

For most purposes, the approximate straight line relationship in Figure 11.7 is adequate, and gives satisfactory results in most cases. The zone diameter ($d$) is also more easily measured than the zone width ($y$).

The BP requires three standard doses in the ratio 1:2:4 to be used, but once linearity has been demonstrated for each compound, a two point (1:2) assay is sufficient. If a dispute arises, a three point assay (1:2:4) must be carried out. The mathematics and statistical analysis of results are discussed by Hewitt.[5]

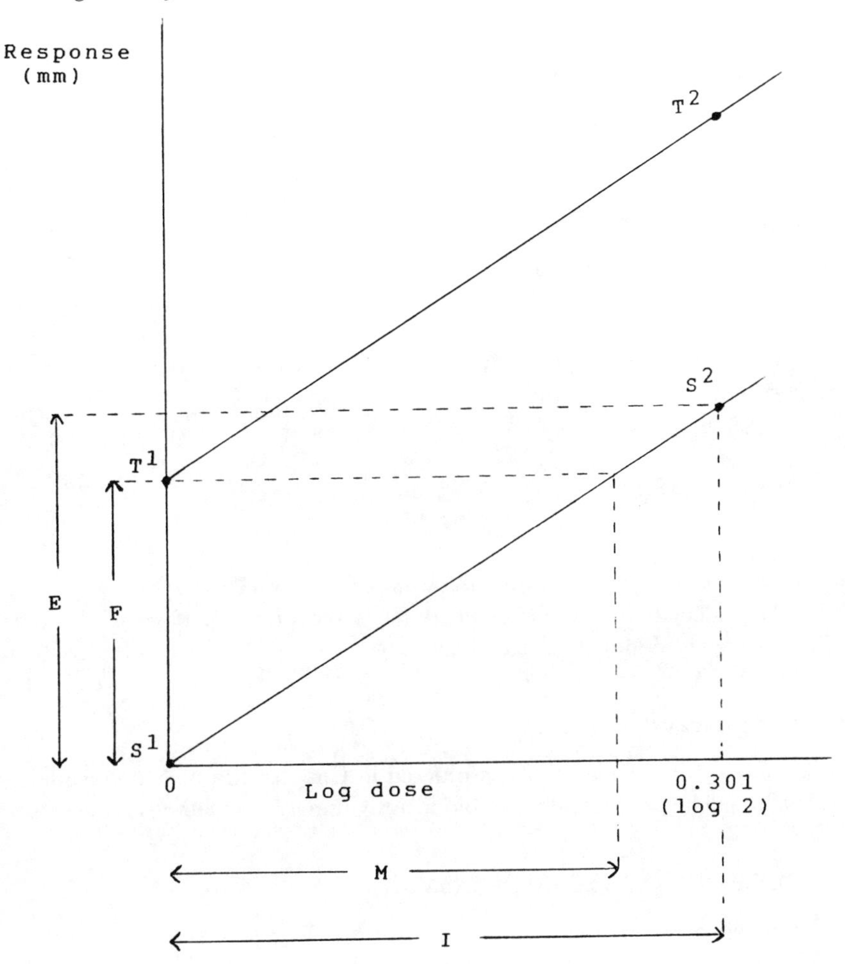


**Figure 11.9** *Example of plot showing how potency ratio is determined*

## 9.4 Calculation of Results

If a 2+2 assay is carried out, *i.e.* two dose levels for one standard and one test at a dose ratio of 1:2, the ideal result is shown in Figure 11.1. Ideally, the lines are parallel and straight and the test solution has the same nominal potency as the standard solution. If this were true, the parallel lines would be the same line, but the presence of two lines shows there is a difference in potency which must be calculated.

In the graph shown (Figure 11.9), $S_1$ and $S_2$ are the responses to the low and high standard doses, $T_1$ and $T_2$ are the responses to the low and high test doses, $M$ is the unknown which has to be determined, and is the log of the ratio of the potencies of the standard and test solutions. $I$ is the log of the dose ratios, *i.e.* log 2 = 0.301 (in this example). $E$ is the difference in response corresponding to a dose increase of 2. $F$ is the difference in response of the organisms to standard and test preparations and corresponds to a difference of $M$ on the log scale. From Figure 11.9, $E$ may be determined as $S_2-S_1$ or $T_2-T_1$. The best estimate of $E$ is given by

$$E=1/2[(S_2-S_1)+(T_2-T_1)]$$

which simplifies to

$$E=1/2[(S_2+T_2)-(S_1+T_1)]$$

Similarly the estimation of $F$ becomes

$$F=1/2[(T_1+T_2)-(S_1+S_2)]$$

From Figure 11.9 the log dose ratio ($I$) corresponds to a response ($E$). From a comparison of similar triangles

$$E/I=F/M$$

and the value for $M$, *i.e.* the log potency ratio is

$$M=FI/E$$

The potency ratio is obtained from the value of antilog $M$.

Similar calculations can be applied for three or more doses and worked examples are given in several references.[2,5]

### 9.5 Tube Assays

Tube assays of disinfectants are considered in Chapter 10 under the heading of MIC values. The principles for antibiotic assay using these methods is the same.

## 10 METHODS OF VITAMIN ASSAY

Microbiological assay of certain members of the 'B' group of vitamins may also be carried out. The micro-organisms used need the vitamin or a derivative of it to form coenzymes, but they are unable to synthesise it in the necessary form.

Agar diffusion methods analogous to those used in antibiotic assays may be carried out, the results being calculated in a similar manner. Tube assays may also be used. The basic principles of vitamin bioassay are very similar to those of antibiotic assay, but there are a number of differences. The organisms are grown in a complete medium which lacks only the compound being assayed. The addition of the missing compound allows microbial growth. Ideally the amount of growth is related to the quantity of vitamin present. There are, however, both theoretical and practical problems.

The vitamin may be a number of chemically related compounds, all allowing growth, but the organisms may respond differentially to each individual compound. Well known examples are the folic acid group and the pyridoxal group of vitamins (vitamin $B_6$) (see Figure 11.10). The most probable explanation is the differing abilities of compounds to pass through the cell membrane, even though they are chemically related. A further problem is that the requirement for the vitamin may not be total, and growth may occur at a reduced rate in its absence.

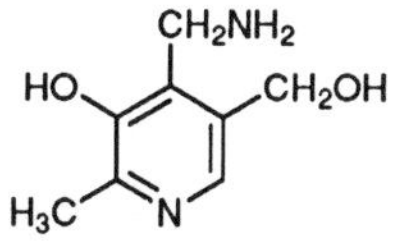

Pyridoxamine

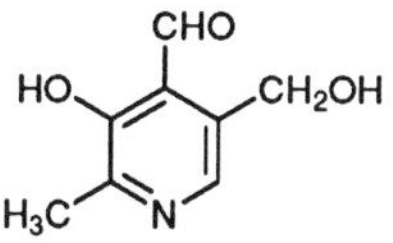

Pyridoxal

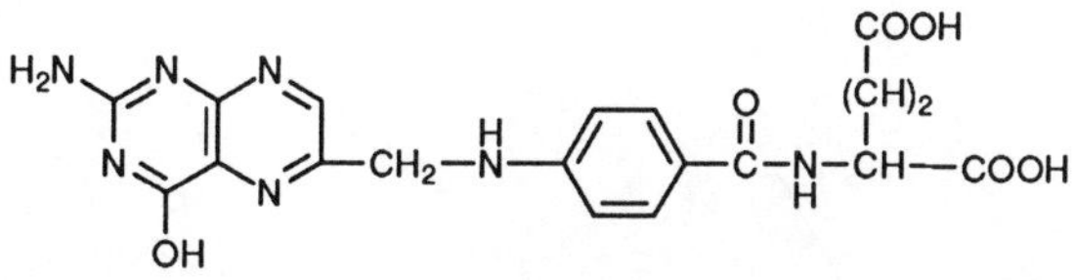

Folic acid

**Figure 11.10** *Folic acid and vitamins of the pyridoxal group*

## 10.1 Agar Diffusion Assays

These are basically the same as the antibiotic diffusion system, except that a zone of growth is formed. A series of high and low standards and tests are prepared and the calculations are identical to those carried out in the antibiotic assays (Figure 11.9).

Generally vitamin assays require a high level of inoculum, and the main precaution is to avoid any carry over of medium with the inoculum, as this may contain vitamin. The assay organism is initially grown on a rich medium containing high levels of vitamin to ensure good growth. To avoid carry over of vitamin, the bacteria are harvested aseptically, either by centrifuging or filtration, washed, and resuspended in sterile saline before use as an inoculum. This process requires considerable microbiological expertise to carry it out satisfactorily.

## 10.2 Tube Assays

These measure growth turbidometrically, or less commonly may use a titrimetric method to measure the level of acid formed during growth. Turbidity may be measured either by means of a nephelometer, if growth has taken place in the appropriate tubes, or spectrophotometrically. The samples are compared with a control which has not been inoculated. The titrimetric method is based on the acid produced during metabolism. The response is complex as the formation of acid may be simultaneous with cell growth, lag behind cell growth, or be a mixture. The theory and mathematics are discussed by Hewitt and Vincent.[2]

There is an excess of vitamin during the early stages of growth, allowing growth to be logarithmic, that is, the logarithm of the number of cells will be linear against time. When the vitamin becomes growth limiting, the growth becomes arithmetic, that is, the number of cells becomes linear with time.

Any attempt to assay the vitamin before all tubes have reached the arithmetic

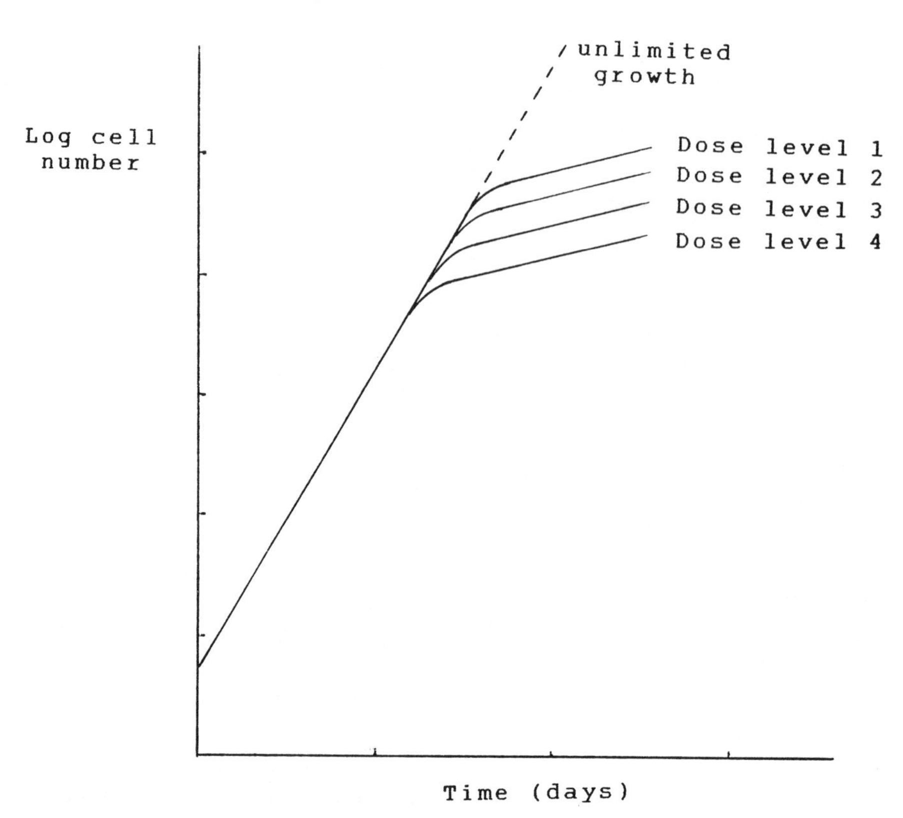


**Figure 11.11** *Non-linear dose-responses arising when the assay is carried out before all the tubes have reached the arithmetic growth stage*

stage of growth, leads to an inadequate response to the higher doses, and produces a non-linear dose-response (Figure 11.11). A minimum time of incubation therefore needs to be determined for each micro-organism/vitamin combination. Hewitt and Vincent[2] consider the theory more fully.

The above paragraph explains why long incubation periods (up to 72 hours) are necessary for these assays. It also explains the necessity for sterile techniques during the assay. Introduction of a contaminant capable of growth in the absence of the vitamin would completely negate the assay.

Heat is frequently used to sterilise or pasteurise the tubes before addition of the inoculum, but causes problems if the vitamin is not heat stable. In this case, membrane filtration may be used, and the vitamin preparation added aseptically to tubes of sterile medium. These procedures again imply a considerable level of microbiological expertise and competence.

It is also necessary to prepare a number of tubes containing high doses of the vitamin. These do not form any part of the assay procedure, but are to ensure that the organism does in fact respond to the vitamin.

As is the case with antibiotics, the microbiological assay of vitamins is time consuming, and a considerable level of microbiological expertise is required. The methods are also staff intensive and are only cost effective if a large number of samples are processed on a regular basis.

## 11 REFERENCES AND FURTHER READING

1. M. Roberts and C.B.C. Boyce, 'Principles of Biological Assay', in 'Methods in Microbiology,' Vol. 7A. ed. J.R. Norris and D.W. Ribbons, Academic Press, London and New York, Ch. 4, pp. 153–190, 1972.
2. W. Hewitt and S. Vincent, 'Theory and Application of Microbiological Assay', Academic Press, San Diego, 1989.
3. 'British Pharmacopoeia', British Pharmacopoeia Commission, HMSO, London, 1980.
4. K.E. Cooper, 'The Theory of Antibiotic Inhibition Zones', in 'Analytical Microbiology', ed. F. Kavanagh, Academic Press, New York and London, Ch. 1, pp. 1–86, 1963.
5. W. Hewitt, 'Microbiological Assay, An Introduction to Quantitative Principles and Evaluation', Academic Press, New York, San Francisco, and London, 1977.
6. J.S. Simpson, Microbiological Assay Using Large Plate Methods', in 'Analytical Microbiology', ed. F. Kavanagh, Academic Press, New York and London, Ch. 2, pp. 88–124, 1963.
7. K.E. Cooper, 1972, 'The Theory of Antibiotic Diffusion Zones', in 'Analytical Microbiology', Vol. II, ed. F. Kavanagh, Academic Press, New York and London, Ch. 2, pp. 13–30, 1972.

## APPENDIX 1

Major suppliers of media within the UK.

Difco Laboratories,
PO Box 14B,
Central Avenue,
East Molesey, East Surrey,
UK.

Oxoid Ltd,
Wade Road,
Basingstoke, Hampshire,
UK.

## APPENDIX 2

Suppliers of pure cultures.

National Collection of Industrial Bacteria (NCIB),
Torey Research Station,
135 Abbey Road,
Aberdeen,
UK.

National Collection of Type Cultures (NCTC)
Central Public Health Laboratory,
Colindale Avenue,
London,
UK.

National Collection of Yeast Cultures (NCYC),
Brewing Research Foundation,
Lyttel Hall,
Nutfield, Redhill,
UK.

CAB International Mycological Institute (IMI),
Alderhurst,
Bakeham Lane,
Egham, Surrey,
UK.

American Type Culture Collection (ATCC),
12301 Parklawn Drive,
Rockville,
Maryland 20852,
USA.

# Subject Index